3rd International Conference on Advances on Clean Energy Research (ICACER 2018)

AA003898

E3S Web of Conferences Volume 51 (2018)

Barcelona, Spain
6 – 8 April 2018

Editors:

W. Cui
E. Rusu

ISBN: 978-1-5108-7007-9

Printed from e-media with permission by:

Curran Associates, Inc.
57 Morehouse Lane
Red Hook, NY 12571

Some format issues inherent in the e-media version may also appear in this print version.

Printed by Curran Associates, Inc. (2018)

For additional information, please contact EDP Sciences – Web of Conferences at the address below.

EDP Sciences – Web of Conferences
17, Avenue du Hoggar
Parc d'Activité de Courtabœuf
BP 112
F-91944 Les Ulis Cedex A
France

Phone: +33 (0) 1 69 18 75 75
Fax: +33 (0) 1 69 28 84 91

contact-edps@webofconferences.org

Additional copies of this publication are available from:

Curran Associates, Inc.
57 Morehouse Lane
Red Hook, NY 12571 USA
Phone: 845-758-0400
Fax: 845-758-2633
Email: curran@proceedings.com
Web: www.proceedings.com

TABLE OF CONTENTS

Assessment of wind energy potential at Kuwaiti Islands by statistical analysis of wind speed data

Khaled Al-Salem[1*], Waleed Al-Nassar[1]
[1]Kuwait Institute for Scientific Research

Abstract. Kuwait possesses a potential of renewable energy, such as solar and wind energy. Wind energy is an alternative clean energy source compared to fossil fuel, which pollute the lower layer of the atmosphere. In this study, statistical methods are used to analyze the wind speed data at Mubarak port (at Bubiyan Island), Failaka Island and Um-AlMaradim Island; which are located respectively in the north, mid and south of Kuwait territorial waters. Wind speed is the most important parameter in the design and study of wind energy conversion systems. The wind speed data were obtained from the Costal Information System Database (CIS) at Kuwait Institute for Scientific Research [1, 2 and 3]over a thirty seven years period, 1979 to 2015. In the present study, the wind energy potential of the locations was statistically analyzed based on wind speed data, over a period of thirty seven years. The probability distributions are derived from the wind data and their distributional parameters are identified. Two probability density functions are fitted to the probability distributions on a yearly basis. The wind energy potential of the locations was studied based on the Weibull and the Rayleigh models.

1 Introduction

Increasing negative effects of fossil fuel combustion on the environment in addition to its limited stock have forced many countries to explore and change to environmentally friendly alternatives that are renewable to sustain the increasing energy demand. Changing to renewable sources and the implementation of effective conservation measures would ensure sustainability. Alternatives to conventional energy sources, especially the renewable ones, are becoming increasingly attractive because of the limited fossil fuel reserves and the adverse effects associated with their use. Some of renewable energy resources are solar and wind. All these renewable energy resources are abundant in Kuwait [4 and 5]. If these resources are well harnessed, they can provide complete security of energy supply. Currently, the wind energy is one of the fastest developing renewable energy source technologies across the globe. Wind energy is an alternative clean energy source compared to fossil fuel, which pollute the lower layer of the atmosphere. It has the advantage of being suitable to be used locally in rural and remote areas. The increasing demand for energy supply coupled with limited energy resources creates an urgency to find new solutions for this energy shortage. Nowadays, wind analysis gives remarkable information to researchers involved in renewable energy studies. Knowledge of the statistical properties of the wind speed is essential for predicting the energy output of a wind energy conversion system. Because of the high variability in space and time of wind energy, it is important to verify that the analyzing method used for the wind data will yield the estimated energy of the wind speed distribution. One of the wind characteristics, is of great importance not only for structural and environmental design and analysis, but also for the assessment of the wind energy potential and the performance of wind energy conversion system as well. Over the last two decades many researches have been devoted to develop an adequate statistical model to describe wind speed frequency distribution. The Weibull, Rayleigh and Lognormal functions are commonly used for fitting the measured wind speed probability distribution [6]. The aim of the present work is to evaluate the potentiality of wind energy at the following locations shown in figure 1:

- Mubarak Port located in Bubiyan island; which is one of the largest island in Kuwait waters and located in the North of Kuwait (29° 50' 31.3"N, 48° 17' 35.3"E).
- Failaka island located in the middle of the terrestrial waters of Kuwait (29° 26' 08.0"N 48° 16' 41.3"EE).
- Umm-AlMaradim island located in the south of terrestrial waters of Kuwait (28° 40' 44.7"N 48° 39' 05.1"E).

*Corresponding author: ksalem@kisr.edu.kw

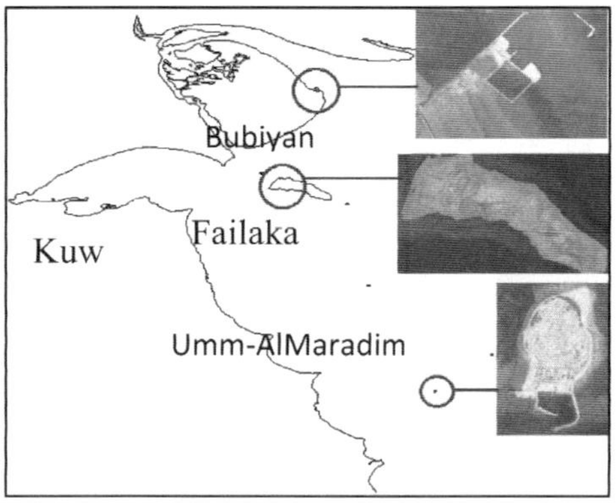

Figure 1. Location of Kuwaiti Islands in Arabian Gulf.

2 Methodology and Analysis

Frequency distribution of wind speed: The wind speed probability density distributions and their functional forms represent the major aspects in wind related literature. Their use includes a wide range of applications, including identifying the parameters of the distribution functions and analyzing the wind speed data as well as wind energy. The wind speed distribution can be determined by dividing the wind speed domain into a number of intervals, mostly of equal width of 1 m/s. The suitable statistical analysis, of the wind speed data in time series format were transformed into frequency distribution format. Based on the wind speed classes, the frequency distribution of the wind speed was established and shown by the Table 1. The wind speed frequency of occurrence of each speed class $f(v_i)$ can found by the following:

$$f(v_i) = \frac{f_i}{N} \quad (1)$$

Where , vi the mean wind speeds calculated for each speed class intervals.

fi the frequency of occurrence of each speed class

N the total data occurrence of each class (12 number of month in a year).

Two of the commonly used functions for fitting a measured wind speed probability distribution in a given location over a certain period of time are the Weibull and Rayleigh distributions [7 and 8]. The probability density function of the Weibull distribution is given as follows:

$$f(v) = \left(\frac{k}{c}\right)\left(\frac{v}{c}\right)^{k-1} exp[-\left(\frac{v}{c}\right)^k] \quad (2)$$

Here, $f(v)$ is the probability of observing wind speed v; c is the Weibull scaling parameter and k is the dimensionless Weibull shape parameter. The corresponding cumulative probability function of the Weibull distribution is given by:

$$F(v) = 1 - exp[-\left(\frac{v}{c}\right)^k] \quad (3)$$

The Rayleigh distribution is a special case of the Weibull distribution in which the shape parameter k takes the value 2.0. From equation 1 the probability density function for the Rayleigh distribution can be simplified as following:

$$f(v) = \left(\frac{2v}{c^2}\right) exp[-\left(\frac{v}{c}\right)^k] \quad (4)$$

The mean value v_m and standard deviation σ of the Weibull distribution [9] can be computed as:

$$v_m = c\neg[1 + \frac{1}{k}] \quad (5)$$

Where standard deviation σ

$$\sigma = c[\neg\left(1 + \frac{2}{k}\right) - \neg^2\left(1 + \frac{1}{k}\right)]^{\frac{1}{2}} \quad (6)$$

Where $\neg$ is the gamma function.

Wind speed variation with height: Wind speed near the ground changes with height. This requires an equation that predicts the wind speed at one height in terms of the measured speed at another. The most common expression for the variation of wind speed with height is the power law [10] having the following form:

$$\frac{v_2}{v_1} = [\frac{h_2}{h_1}]^{\propto} \quad (7)$$

Where v_1 and v_2 are the mean wind speeds at heights h_1 and h_2 , respectively. The exponent $\propto$ depends on factors such as surface roughness and atmospheric stability definitions according to the [European Wind Atlas, WASP]. Numerically, it lies in the range 0.0–0.5, with the most frequently adopted value being 0.5 which is applicable to low surfaces and well exposed will used a roughness length 0.0024.

Where equation 6 can be written as follows:

$$\frac{v_2}{v_1} = \frac{\ln(\frac{h2}{r})}{\ln(\frac{h1}{r})} \quad (8)$$

Where r is Roughness Length (m), v_2 wind speeds at selected ground level h_2.

The wind speed at a certain height above ground level is

$$v_2 = v_1[\frac{\ln(\frac{h2}{r})}{\ln(\frac{h1}{r})}] \quad (9)$$

The fact that the wind profile is twisted towards a lower speed as we move closer to ground level is usually called wind shear. Wind shear may also be important when designing wind turbines.

Wind Power density function: It is well known that the power of the wind at speed v through a wind turbines blade sweep area (A) increases as the cube of its velocity and can be calculated as follows [2 and 11] :

$$P(v) = \frac{1}{2}\rho A v^3 \quad (10)$$

Where ρ is the mean air density (1.225 $\frac{kg}{m^3}$ at average atmospheric pressure at sea level and at 15° C). The expected monthly or annual wind power density per unit area of a site based on a Weibull probability density function can be expressed as follows:

$$P_w = \frac{1}{2}\rho c^3\neg[1 + \frac{3}{k}] \quad (11)$$

To calculate Weibull shape parameter or factor (K) is by iteration method by using maximum likelihood method as the following:

$$k = \left(\frac{\sum_1^n v^k Ln(v)}{\sum_1^n v^k} - \frac{\sum_1^n Ln(v)}{n}\right)^{-1}$$

Where c is the Weibull scale parameter (m/s), and is given by:

$$c = \frac{v_m}{\neg(1+\frac{1}{k})} \quad (12)$$

Then, Weibull scale parameter is calculated as follows

$$c = (\frac{\sum_1^n v^k}{n})^{\frac{1}{k}}$$

The two significant parameters k and C are closely related to the mean value of the wind speed v_m [12]. By extracting C from Equation [12] and K, the power density for the Rayleigh model can be found by [13]:

$$P_R = \frac{3}{\pi}\rho c^3 (\frac{\pi}{4})^{\frac{3}{2}} \quad (13)$$

Errors in calculating the power densities using the distributions in comparison to values of the probability density distributions derived from measured values can be found by the following:

$$Error(\%) = \frac{(P_{W,R} - P_{m,R})}{P_{m,R}} \quad (14)$$

where $P_{W,R}$ is the mean power density calculated from either the Weibull or Rayleigh function used in the calculation of the error and $P_{m,R}$ is the wind power density for the probability density distribution, derived from measured values which serves as 'the reference mean power density'. And can be calculated from the following [13]:

$$P_{m,R} = \sum_{j=1}^{n} [\frac{1}{2}\rho v_m^3 f(v_j)] \quad (15)$$

The yearly average error value in calculating the power density using the Weibull function is given by the following:

$$Error(\%) = \frac{1}{12}\sum_{i=1}^{12} [\frac{(P_{W,r} - P_{m,R})}{P_{m,R}}] \quad (16)$$

The statistical analysis of the distributions: The square of the correlation coefficient (R^2), chi-square (X^2) and root mean square error analysis (RMSE) are used to evaluate the performances of the Weibull and Rayleigh distributions [6]. These parameters can be calculated as follows:

$$R^2 = \frac{(\sum_{i=1}^{N}(y_i - Z_i)^2 - \sum_{i=1}^{N}(x_i - y_i)^2)}{\sum_{i=1}^{N}(y_i - Z_i)^2} \quad (17)$$

$$x^2 = \frac{\sum_{i=1}^{n}(y_i - x_i)^2}{N - n} \quad (18)$$

$$RMSE = [\frac{1}{N}\sum_{i=1}^{N}(y_i - x_i)^2]^{\frac{1}{2}} \quad (19)$$

Where y_i is the measured data, Z_i is the mean value, x_i is the predicted data with the Weibull or Rayleigh distribution, N is the number of observations and n is the number of constants. Therefore, the distribution function can be selected according to the highest value of R^2 and the lowest values of RMSE and x^2.

3 Discussion and Results

Data for wind speed used in the present calculations were obtained during the period 1979–2015 from the Coastal Information system Database at KISR[1, 2 and 3]. The wind data used were numerically predicted and validated with Kuwait Airport data[2 and 14] and the three locations in the study were positioned in open spaces free of obstacles at a height of 30m above the ground. Wind speeds stored and averaged over one hour in the database. The main results obtained from the present study can be summarized as follows.

The Monthly mean wind speed values at 30m height and the standard deviations calculated from the available data for period of 37 years (1979-2015) as shown in Table 1 for the three selected location presented as follows:

- For Mubarak port at Bubiyan Island indicated for the overall 37 years, the mean wind speeds is about 4.5 m/s. The wind speed for the whole year has the maximum monthly mean value of 5.4 m/s which arises in June, while a minimum value of 3.9 m/s occurs in October.
- Similarly, for Failaka Island, the mean wind speeds is about 4.46 m/s. The wind speed for the whole year has the maximum monthly mean value of 5.24 m/s which arises in June, while a minimum value of 3.77 m/s occurs in October.
- Similarly, for Umm AlMaradim Island, the mean wind speeds is about 4.66 m/s. The wind speed for the whole year has the maximum monthly mean value of 5.44 m/s which arises in June, while a minimum value of 3.97 m/s occurs in October.

To construct the wind speed probability distributions we need to determine frequency distribution of the wind speed, we must first divide the wind speed domain into a number of intervals, mostly of equal width of 1 m/s. As a result, for a suitable statistical analysis, the wind speed data in time series format were transformed into monthly frequency distribution format. Based on the wind speed bins, the frequency distribution of the wind speed data was established for the three selected location and shown by Tables 2,3 and 4. As the tables indicates, the trends of the monthly means for the different years are similar for all selected locations.

The variation of wind velocity is often described using the Weibull two-parameter density function. This is a statistical method which is widely accepted for evaluating local wind load probabilities and considered as a standard approach. The Weibull shape function k and scale factor c parameters are calculated analytically from the available data are presented as a mean values in Table 5 for Bubiyan Island, Failaka Island and for Umm AlMaradim. As shown, the scale parameter c for the selected locations varies over the whole period (1979-2015) show the following:

- Bubiyan Island 1.63 (November 2009) and 6.45 m/s (August 1994).
- Failaka Island 1.9 (October 2008 and December 2009) and 6.7 m/s (June 2005).
- Umm AlMaradim Island 1.38 (December 2009) and 6.47 m/s (July 1994).

Where shape parameter K ranges as follows:

- Bubiyan Island 1.42 (November 2009) and 4.21 (July 1994).
- Failaka Island 1.45 (February 2009) and 4.13 (August 1997).
- Umm AlMaradim Island 1.34 (January 2009) and 4.07 (July 1991).

https://doi.org/10.1051/e3sconf/20185101001

Table 1. Monthly mean wind speeds (m/s) and standard deviations (σ) at Bubiyan, Failaka and Umm AlMaradim Islands in Kuwait waters, for periods of 1979–2015

Locations		Jan	Feb	Mar	Apr	May	Jun	Jul	Aug	Sep	Oct	Nov	Dec	Year
Bubiyan Island	M W.S	4.46	4.81	4.77	4.67	4.67	5.39	5.21	4.62	4.08	3.90	4.25	4.34	4.60
	σ	1.99	2.13	2.15	2.04	1.98	2.31	2.11	1.91	1.79	1.70	1.82	1.90	1.99
Failaka Island	M W.S	4.29	4.64	4.63	4.47	4.45	5.24	5.16	4.58	4.01	3.77	4.11	4.22	4.46
	σ	1.89	2.02	2.06	1.98	1.93	2.24	2.08	1.89	1.78	1.63	1.74	1.82	1.92
Umm AlMaradim	M W.S	5.54	5.81	5.68	5.50	5.50	6.47	6.21	5.51	4.90	4.70	5.27	5.43	5.54
	σ	2.40	2.53	2.56	2.39	2.34	2.79	2.57	2.33	2.18	2.05	2.25	2.34	2.39

Table 2. Wind speed data transformed into monthly frequency distribution format for Mubarak Port at Bubiyan Island at 30m height for periods 1979-2015.

	Wind Speed bins (m/s)										
No/Sp	0.-1.0	1.-2.0	2.-3.0	3.-4.0	4.-5.0	5.-6.0	6.-7.0	7.-8.0	8.-9.0	9.-10.	>10
Jan	0.029	0.0863	0.1368	0.1961	0.1889	0.1484	0.0933	0.0616	0.0343	0.0145	0.0109
Feb	0.0166	0.0644	0.1222	0.199	0.1871	0.1449	0.1031	0.0743	0.0409	0.0257	0.0218
Mar	0.0203	0.067	0.1332	0.1902	0.1817	0.1411	0.1044	0.0719	0.0453	0.0223	0.0226
Apr	0.0203	0.0695	0.1331	0.1887	0.1914	0.151	0.1056	0.0655	0.0405	0.0206	0.0136
May	0.0186	0.0689	0.126	0.1949	0.1961	0.1543	0.103	0.0663	0.039	0.0191	0.0139
Jun	0.022	0.0609	0.0974	0.1369	0.1505	0.1449	0.1274	0.1023	0.0702	0.0445	0.0431
Jul	0.0243	0.0574	0.1015	0.1434	0.1713	0.1524	0.1218	0.0967	0.0638	0.0386	0.0289
Aug	0.0282	0.0776	0.1309	0.1796	0.1977	0.1391	0.1003	0.0695	0.0441	0.0196	0.0134
Sep	0.0304	0.1042	0.1668	0.2176	0.2117	0.1168	0.072	0.0428	0.0221	0.0098	0.006
Oct	0.0368	0.1095	0.1846	0.2435	0.1914	0.1043	0.0655	0.0336	0.0159	0.0068	0.0082
Nov	0.0338	0.0931	0.1473	0.2163	0.1867	0.1392	0.0938	0.0474	0.023	0.0114	0.008
Dec	0.0288	0.0925	0.1529	0.2077	0.1869	0.1333	0.0852	0.0556	0.0281	0.0169	0.0121
Year	0.0258	0.0793	0.1361	0.1928	0.1868	0.1391	0.0979	0.0656	0.0389	0.0208	0.0169
Cumulative	0.0258	0.105	0.2411	0.4339	0.6207	0.7598	0.8578	0.9234	0.9623	0.9831	1

Table 3. Wind speed data transformed into monthly frequency distribution format for Failaka Island at 30m height for periods 1979-2015.

	Wind Speed bins (m/s)										
No/Sp	0.-1.0	1.-2.0	2.-3.0	3.-4.0	4.-5.0	5.-6.0	6.-7.0	7.-8.0	8.-9.0	9.-10.	>10
Jan	0.0295	0.0882	0.1453	0.2278	0.183	0.1395	0.0925	0.0488	0.025	0.0119	0.0086
Feb	0.0218	0.0638	0.1255	0.2181	0.1933	0.1455	0.0998	0.0594	0.0347	0.0221	0.0161
Mar	0.0234	0.0682	0.1281	0.2035	0.1866	0.1534	0.1073	0.0632	0.0317	0.0178	0.0168
Apr	0.0269	0.0774	0.1413	0.2044	0.1884	0.1481	0.0939	0.0588	0.0352	0.0148	0.0108
May	0.0209	0.0786	0.1458	0.2099	0.1988	0.1355	0.096	0.0604	0.0288	0.0156	0.0098
Jun	0.0188	0.0654	0.1052	0.1572	0.1555	0.1445	0.1171	0.09	0.0653	0.0404	0.0405
Jul	0.0181	0.0616	0.1045	0.1548	0.1725	0.1512	0.1205	0.0913	0.0624	0.0348	0.0283
Aug	0.0231	0.0767	0.1392	0.1968	0.19	0.1409	0.0911	0.0642	0.0453	0.0184	0.0143
Sep	0.0324	0.1033	0.1859	0.2322	0.1805	0.1142	0.0725	0.0385	0.0231	0.0116	0.0057
Oct	0.0378	0.1138	0.1952	0.2624	0.1829	0.1024	0.0511	0.0279	0.0139	0.0058	0.0068
Nov	0.0302	0.0981	0.1615	0.2418	0.1912	0.1185	0.0738	0.0468	0.0222	0.0083	0.0077
Dec	0.0353	0.0893	0.152	0.2318	0.1876	0.1277	0.0825	0.0463	0.0252	0.0128	0.0096
Year	0.0265	0.082	0.1441	0.2117	0.1842	0.1351	0.0915	0.058	0.0344	0.0178	0.0146
Cumulative	0.0265	0.1085	0.2527	0.4644	0.6486	0.7837	0.8752	0.9332	0.9676	0.9854	1

Table 4. Wind speed data transformed into monthly frequency distribution format for Umm AlMaradim Island at 30m height for periods 1979-2015.

	Wind Speed bins (m/s)										
No/Sp	0.-1.0	1.-2.0	2.-3.0	3.-4.0	4.-5.0	5.-6.0	6.-7.0	7.-8.0	8.-9.0	9.-10.	>10
Jan	0.0363	0.0851	0.1223	0.1718	0.1783	0.1461	0.112	0.0741	0.0404	0.0184	0.0151
Feb	0.0278	0.0641	0.1147	0.1722	0.183	0.1499	0.1105	0.0783	0.0486	0.0286	0.0222
Mar	0.032	0.0697	0.1221	0.1756	0.1767	0.1454	0.1078	0.0761	0.0466	0.024	0.0241
Apr	0.028	0.0716	0.1291	0.1783	0.1875	0.1533	0.1071	0.0668	0.0416	0.0221	0.0148
May	0.0256	0.0744	0.1203	0.1862	0.1925	0.1522	0.1019	0.0703	0.0422	0.0197	0.0146
Jun	0.0249	0.0632	0.0999	0.13	0.1449	0.1391	0.1193	0.1043	0.0789	0.05	0.0456
Jul	0.0295	0.0609	0.1043	0.1418	0.1574	0.1447	0.1222	0.0979	0.066	0.0425	0.0328
Aug	0.0356	0.0865	0.1301	0.177	0.1765	0.1395	0.0964	0.0724	0.046	0.0255	0.0144
Sep	0.0409	0.1015	0.167	0.2073	0.1989	0.1134	0.0801	0.0458	0.026	0.0117	0.0074
Oct	0.0476	0.1011	0.1751	0.2292	0.1881	0.1133	0.0693	0.0393	0.0194	0.0089	0.0085
Nov	0.0427	0.0921	0.1358	0.1856	0.1834	0.1357	0.1011	0.0636	0.0316	0.0177	0.0106
Dec	0.0439	0.0886	0.1313	0.18	0.1782	0.1381	0.1031	0.0657	0.0357	0.0189	0.0165
Year	0.0346	0.0799	0.1293	0.1779	0.1788	0.1392	0.1026	0.0712	0.0436	0.024	0.0189
Cumulative	0.0346	0.1145	0.2438	0.4217	0.6005	0.7397	0.8423	0.9135	0.9571	0.9811	1

Table 5. Monthly Weibull shape parameters k and scale parameters c for at Bubiyan, Failaka and Umm AlMaradim Islands for periods of 1979–2015.

	Par	Jan	Feb	Mar	Apr	May	Jun	Jul	Aug	Sep	Oct	Nov	Dec	Year
Bubiyan Island	K	2.42	2.47	2.37	2.47	2.54	2.53	2.70	2.61	2.49	2.53	2.52	2.48	2.51
	C	3.96	4.28	4.23	4.14	4.15	4.80	4.65	4.11	3.63	3.47	3.78	3.86	4.09
Failaka Island	K	2.46	2.53	2.42	2.46	2.52	2.52	2.70	2.63	2.44	2.54	2.57	2.55	2.53
	C	3.81	4.13	4.11	3.97	3.95	4.66	4.59	4.08	3.56	3.35	3.65	3.76	3.97
Umm AlMaradim Island	K	2.50	2.51	2.38	2.49	2.53	2.53	2.68	2.58	2.47	2.55	2.54	2.53	2.52
	C	4.92	5.16	5.04	4.88	4.88	5.75	5.54	4.90	4.36	4.18	4.69	4.83	4.93

In order to observe the Weibull distribution of the three selected location, the Weibull probability density distributions for each of the 37 years were analyzed. The distributions obtained for the three locations are illustrated in figure 2 which represents Bubiyan island, Failaka Island and Umm AlMaradim Island respectively. It can be seen that the distributions for the three locations are similar for the 37 years period and represents a maximum peak at wind speed of about 4.0 m/s. The Rayleigh approximations of the actual probability density distribution of wind speeds for the period (1979-2015) for the three selected location are shown in the same figure. A comparison of the actual probability distribution of wind speeds derived from CIS database with Weibull and Rayleigh approximations for whole year presented in Table 6 for Bubiyan, Failaka and Umm AlMaradim islands respectively. It can be seen that the distributions for the three locations are similar for the 37 years period.

Table 6. Comparison of the actual (CIS data) probability distribution of wind speeds derived from data with Weibull and Rayleigh for periods 1979-2015.

			Bubiyan Island			Failaka Island			Umm AlMaradim Island		
j	v_j	v_{mj}	CIS	Weibull	Rayleigh	CIS	Weibull	Rayleigh	CIS	Weibull	Rayleigh
1	0-1	0.5	0.026	0.034	0.062	0.027	0.033	0.066	0.035	0.042	0.064
2	1-2	1.5	0.079	0.128	0.169	0.082	0.136	0.179	0.080	0.126	0.167
3	2-3	2.5	0.136	0.212	0.223	0.144	0.228	0.233	0.129	0.201	0.217
4	3-4	3.5	0.193	0.239	0.210	0.212	0.249	0.210	0.178	0.229	0.206
5	4-5	4.5	0.187	0.195	0.149	0.184	0.189	0.140	0.179	0.194	0.152

6	5-6	5.5	0.139	0.117	0.081	0.135	0.103	0.070	0.139	0.122	0.087
7	6-7	6.5	0.098	0.052	0.034	0.092	0.043	0.029	0.103	0.058	0.039
8	7-8	7.5	0.066	0.018	0.012	0.058	0.015	0.010	0.071	0.022	0.015
9	8-9	8.5	0.039	0.005	0.004	0.034	0.005	0.003	0.044	0.007	0.005
10	9-10	9.5	0.021	0.002	0.001	0.018	0.001	0.001	0.024	0.002	0.002
11	>10	>10	0.017	0.001	0.001	0.015	0.001	0.001	0.019	0.001	0.001

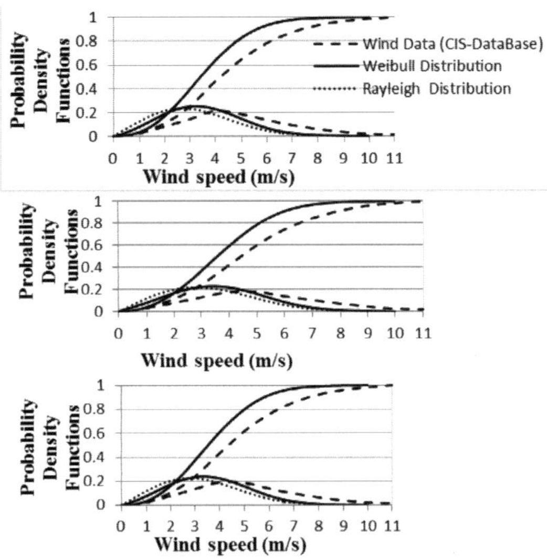

Figure 2. Wind speed probability density from Weibull and Reyleigh distribution and cumulative probability distributions for periods if 1979-2015 At Bubiyan, Failaka and Umm ALMaradim Islands respectively.

The parameters for the statistical analysis correlation coefficient R^2, the root mean square error, $RMSE$, and Chi-square error X^2, are given in Table 7 for the three selected locations. As it can be seen in Table 7 the value of correlation coefficient for Weibull is higher than the Rayleigh. The correlation coefficient values range from 0.6223 to 0.509 for the Weibull distribution, and Rayleigh distributions from 0.31 to 0.13 for the three selected location. However, the results have shown that the RMSE and X^2 values of the Weibull distribution are lower than the values obtained by the Rayleigh distribution for the three selected location. As result, the Weibull approximation is found to be the most accurate distribution according to the lower values of RMSE and higher R^2.

The frequency of wind power densities calculated from CIS database probability distributions are shown in figures 3, 4 and 5 and Tables 8, 9 and 10 for the three selected locations. The power density shows in percent of Frequency. The average power densities occur at each location as follows:

- For Bubiyan Island the highest yearly power density ranges of 150 w/m^2 for frequency of occurrences about 36% and a cumulative WPD of 66%.
- For Failaka Island the highest yearly power density range of 150 to 200 w/m^2 for frequency of occurrences about 32% and cumulative WPD of 58%.
- For Umm AlMaradim Island the highest yearly power density ranges of 150 to 200 w/m^2 for frequency of occurrences about 29% and cumulative WPD of 55%.

Design performance from a wind turbine, the wind speed at the three selected locations is the most important factor in how well a wind turbine performs for the amount of wind that's available. It's of little use installing a large turbine in an area which only receives a light breeze and likewise, it's a waste to install a smaller system on a site which could easily allow for greater power generation. The design performance should follow important factors as follows:

- A turbine's cut-in and cut-out speed are determined by the manufacturer to protect the turbine from damage. The cut-in speed is simple; it's the point at which the turbine starts generating electricity from turning. The cut-out point is more important, and denotes how fast the turbine can go before wind speeds get so fast that it risks damage.
- Rated peak speed - that is, a wind speed at which its return the optimal amount of power.

Figure 3, 4 and 5 are shows the design performance factors for a wind turbine at the three selected location. The cut-in speed which is at about 3 m/s and the cut-out will be at about more the 10 m/s and the peek speed for optimal of power generation 5-6 m/s at wind power 175 w/m^2.

Table 7. The statistical analysis parameters for wind speed distributions for Period of 1979-2015.

	Bubiyan Island		Failaka Island		Umm AlMaradim Island	
	Weibull	Rayleigh	Weibull	Rayleigh	Weibull	Rayleigh
R^2	0.5936	0.2642	0.6223	0.3131	0.5099	0.1378
X^2	0.0019	0.0034	0.0020	0.0036	0.0019	0.0033
$RMSE$	0.0394	0.0530	0.0405	0.0546	0.0394	0.0522

4 CONCLUSIONS

The objective of this study was to investigate the potential of wind energy resource in Mubarak Port area at Bubiyan island, Failaka Failaka Island and Umm AlMaradim Island in Kuwait. For this purpose, wind speed data obtained from the Costal Information System (CIS) database at the Kuwait institute for Scientific Research (KISR) were analyzed over a 37 years period from 1979 to 2015. The probability density distributions and power density were derived from the time series data. Weibull and Rayleigh probability density function have been fitted to the wind speed and the wind power density was evaluated. The most important outcomes of the study can be summarized as follows:

- The predicted average annual peak wind power density for Bubiyan Island, Failaka Island and a Umm AlMaradim Island are 150, 175 and 175 w/m^2, respectively. These peak wind power is available for about 36%, 32% and 29% of the year for these locations.
- For Boubyan, Failaka and Umm Al-Maradim Islands, wind power exceeds 150 w/m^2 for 33.6%, 42.3% and 56.5% in a year respectively

The Weibull distribution is better than the Rayleigh distribution for the monthly probability density distribution for the whole years. All the three sites fall under wind power class 2 (As per Pacific Northwest Laboratory (PNL) classification), since the wind power density values are more than 100 w/m^2. Hence all these three sites are ideal for grid-connected applications.

Table 8. Frequency Distribution of Wind Power Density (%) for Bubiyan Island (1979-2015).

Wind Power Density (W/m^2) Class

No/Sp	0	50	100	150	200	250	300	350	400	450	500	550
Jan	0	0.0	37.8	51.4	10.8	0.0	0.0	0.0	0.0	0.0	0.0	0.0
Feb	0	0.0	8.1	56.8	24.3	8.1	0.0	0.0	2.7	0.0	0.0	0.0
Mar	0	0.0	5.4	51.4	35.1	8.1	0.0	0.0	0.0	0.0	0.0	0.0
Apr	0	2.7	18.9	54.1	16.2	8.1	0.0	0.0	0.0	0.0	0.0	0.0
May	0	0.0	27.0	45.9	16.2	8.1	2.7	0.0	0.0	0.0	0.0	0.0
Jun	0	0.0	2.7	10.8	43.2	10.8	8.1	13.5	2.7	5.4	2.7	0.0
Jul	0	0.0	2.7	29.7	24.3	21.6	5.4	2.7	8.1	5.4	0.0	0.0
Aug	0	0.0	24.3	40.5	13.5	16.2	0.0	0.0	2.7	2.7	0.0	0.0
Sep	0	10.8	35.1	35.1	16.2	2.7	0.0	0.0	0.0	0.0	0.0	0.0
Oct	0	18.9	59.5	10.8	5.4	2.7	0.0	0.0	2.7	0.0	0.0	0.0
Nov	0	5.4	54.1	16.2	13.5	10.8	0.0	0.0	0.0	0.0	0.0	0.0
Dec	0	2.7	51.4	27.0	5.4	8.1	5.4	0.0	0.0	0.0	0.0	0.0
Year	0	3.4	27.3	35.8	18.7	8.8	1.8	1.4	1.6	1.1	0.2	0.0
Cumulative	0	3.4	30.6	66.4	85.1	93.9	95.7	97.1	98.6	99.8	100.0	100

Table 9. Frequency Distribution of Wind Power Density (%) for Failaka Island (1979-2015).

Wind Power Density (W/m^2) Class

No/Sp	0	50	100	150	200	250	300	350	400	450	500	550
Jan	0	0.0	18.9	56.8	8.1	16.2	0.0	0.0	0.0	0.0	0.0	0.0
Feb	0	0.0	5.4	45.9	27.0	16.2	2.7	0.0	2.7	0.0	0.0	0.0
Mar	0	0.0	0.0	43.2	35.1	21.6	0.0	0.0	0.0	0.0	0.0	0.0
Apr	0	2.7	16.2	32.4	40.5	5.4	2.7	0.0	0.0	0.0	0.0	0.0
May	0	0.0	18.9	51.4	16.2	5.4	8.1	0.0	0.0	0.0	0.0	0.0
Jun	0	0.0	0.0	10.8	37.8	21.6	0.0	16.2	2.7	2.7	5.4	2.7

Jul	0	0.0	0.0	24.3	27.0	16.2	13.5	0.0	10.8	2.7	5.4	0.0
Aug	0	0.0	21.6	32.4	21.6	13.5	5.4	0.0	2.7	0.0	2.7	0.0
Sep	0	10.8	37.8	27.0	13.5	10.8	0.0	0.0	0.0	0.0	0.0	0.0
Oct	0	18.9	59.5	8.1	5.4	5.4	0.0	0.0	2.7	0.0	0.0	0.0
Nov	0	5.4	45.9	21.6	13.5	2.7	8.1	2.7	0.0	0.0	0.0	0.0
Dec	0	0.0	40.5	35.1	10.8	0.0	5.4	8.1	0.0	0.0	0.0	0.0
Year	0	3.2	22.1	32.4	21.4	11.3	3.8	2.3	1.8	0.5	1.1	0.2
Cumulative	0	3.2	25.2	57.7	79.1	90.3	94.1	96.4	98.2	98.6	99.8	100

Table 10. Frequency Distribution of Wind Power Density (%) for Umm AlMaradim Island (1979-2015).

	Wind Power Density (W/m^2) Class											
No/Sp	0	50	100	150	200	250	300	350	400	450	500	550
Jan	0	0.0	8.1	43.2	21.6	10.8	10.8	5.4	0.0	0.0	0.0	0.0
Feb	0	0.0	0.0	16.2	45.9	21.6	10.8	2.7	2.7	0.0	0.0	0.0
Mar	0	0.0	0.0	21.6	43.2	27.0	8.1	0.0	0.0	0.0	0.0	0.0
Apr	0	0.0	8.1	29.7	40.5	13.5	8.1	0.0	0.0	0.0	0.0	0.0
May	0	0.0	10.8	35.1	32.4	10.8	10.8	0.0	0.0	0.0	0.0	0.0
Jun	0	0.0	0.0	2.7	18.9	27.0	18.9	21.6	0.0	5.4	2.7	2.7
Jul	0	0.0	2.7	10.8	29.7	16.2	18.9	5.4	10.8	2.7	2.7	0.0
Aug	0	0.0	16.2	29.7	18.9	24.3	5.4	0.0	2.7	2.7	0.0	0.0
Sep	0	0.0	35.1	43.2	13.5	8.1	0.0	0.0	0.0	0.0	0.0	0.0
Oct	0	2.7	56.8	18.9	10.8	8.1	2.7	0.0	0.0	0.0	0.0	0.0
Nov	0	0.0	27.0	37.8	10.8	8.1	8.1	2.7	0.0	5.4	0.0	0.0
Dec	0	0.0	8.1	56.8	18.9	5.4	2.7	0.0	0.0	0.0	8.1	0.0
Year	0	0.2	14.4	28.8	25.5	15.1	8.8	3.2	1.4	1.4	1.1	0.2
Cumulative	0	0.2	14.6	43.5	68.9	84.0	92.8	95.9	97.3	98.6	99.8	100

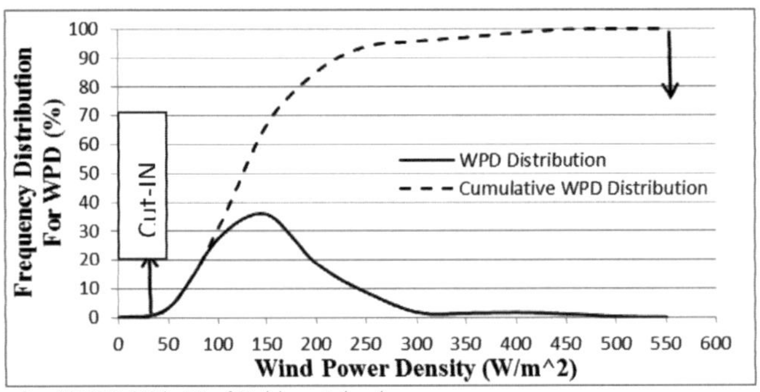

Figure 3. for wind power density (WPD) (W/m2) of Bubiyan Island.

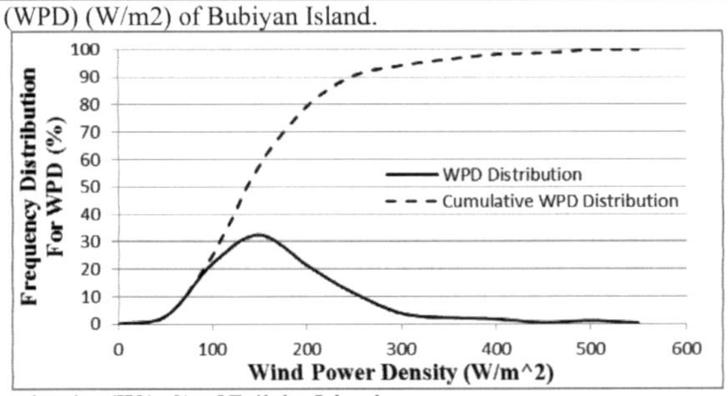

Figure 4. Distribution Wind power density (W/m2) of Failaka Island.

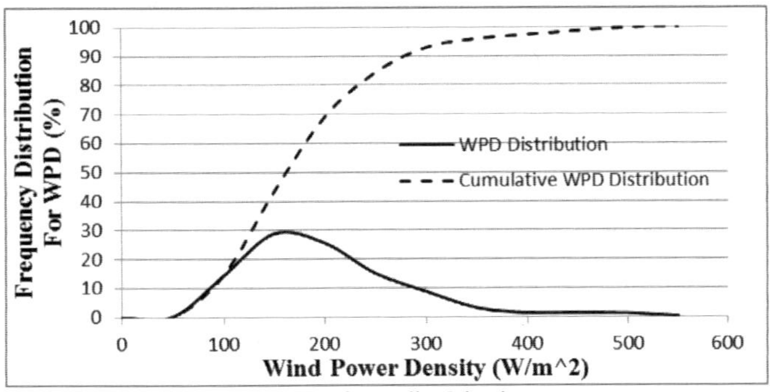

Figure 5. Distribution Wind power density (W/m2) of Umm AlMaradim Island.

REFERENCE

1. Al-Salem K. (2008), "Interactive Coastal Information System for Kuwait's Territorial Waters Phase II: Website site for online Hydrodynamic Predictions". Kuwait Institute for Scientific Research. Report KISR 9485 Kuwait 2008.
2. Al-Salem K. and Al-Rashed A., 2016. "Updating the Database of KISR's Coastal Information System (CIS) by Using Numerical Techniques". Kuwait Institute for Scientific Research. Report KISR 13793. Kuwait 2016
3. Rakha K. and Al-Salem K., (2007) "Interactive Coastal Information System for Kuwait's Territorial Waters Phase I: Hindcast of wave, water levels and current". Kuwait Institute for Scientific Research (EC026C). Kuwait 2007.
4. Al-Naser, W. E.; Al-Naser, N. W. The Status of Renewable Energy in the GCC Countries. Renewable and Sustainable Energy Reviews. 2011, 15 (6), 3074-3098.
5. Al-Nassar. W.; Alhajraf S. A.; Al-Enizi, A.; Al-Awadhi, L. Potential Wind Power Generation in the State of Kuwait. Renewable Energy. 2005, 30, 2149–2161.
6. Meishen Li, Xianguo Li. (2005). MEP-type distribution function: a better alternative to Weibull function for wind speed distributions. Renewable Energy; 30:1221–1240.
7. Egbert Boeker, Rienk Van Grondelle (1999). Environmental Physics. Second edition. John Wiley & SONS, LTD.
8. Ramrez P, Carta JA. (2005). The use of wind probability distributions derived from the maximum entropy principle in the analysis of wind energy. A case study. Energy Conversion and Management; 1-14
9. Celik AN. (2003). A statistical analysis of wind power density based on the Weibull and Rayleigh models at the southern region of Turkey. Renewable Energy; 29:593-604.
10. Perez IA, Garcia MA, Sanchez ML, Torre de B. (2004). Analysis of height variations of sodar-derived wind speeds in Northern Spain. Journal of Wind Engineering and Industrial Aerodynamics; 92:875–894.
11. Ahmed Shata AS, Hanitsch R. (2005). Evaluation of wind energy potential and electricity generation on the coast of editerranean Sea in Egypt. Renewable Energy; 1–20.
12. Celik AN. (2004). On the distributional parameters used in assessment of the suitability of wind speed probability density functions. Energy Conversion and Management; 45: 1735–1747.
13. Al-Mohamad A, Karmeh. (2003): Wind energy potential in Syria. Renewable Energy ; 28: 1039-1046.
14. Al-Salem K., Rakha K. 2005."Verification of a WAM Model for the Arabian Gulf". Submitted to Arabian Coast 2005 Conference at Dubai 15 October 2005.

Biogas Generation Using Kitchen Waste

T Younas[1,*], M Taha [2], S F Ehtesham[2], M F Siddiqui[2]

[1]Assistant Professor, Department of Mechatronics Engineering, SZABIST, Karachi, Pakistan
[2]Students, Department of Mechatronics Engineering, SZABIST, Karachi, Pakistan

Abstract. The previous years has been very crucial for the whole world so in Pakistan. This situation arise due to shocking increment in the rates of oil. In order to overcome this issue most of the countries are working for the development of technology using renewable resources. These resources include solar, wind and biomass. Biomass includes cow dung, kitchen waste, wood etc. The geographical location of Pakistan is a best suit for biomass energy operation. Among these biomasses this paper will be focusing on the kitchen waste which will result in around 60% of methane gas, 30% will include carbon dioxide, 8% nitrogen and rest 1 to 2 % of hydrogen sulphide. This paper will state the best possible option to perform anaerobic digestion process in order to generate excess amount of biogas at homes. It will also discuss procedure for the removal of toxic gases which exist in biogas and can be harmful for humans as well as it degrade biogas quality. In our research, the generation of biogas and methane is done from the sugary and starch-rich material and is determined at small scale using the elementary digesters.

1 Introduction

Kitchen waste is bio-material having the high energy value and calorific value to microorganisms, that's why efficiency of biogas generation can be expanded by several ways of magnitude. The dumping of food waste is the main cause of pollution can be finished in low cost and environment friendly way [1] [2]. Considering cost effective is not the target the dumping of waste must be done in proper way for maintaining hygiene. The disposal of food in areas and making the places unhygienic and it needs the solution. Installation of biogas plants give the proper way of utilizing and proper storage of waste [3]. .The process of anaerobic digestion is a key path that allows to efficiently handle a use biofuel (which is comprised of mostly methane and one fourth of carbon dioxide) in order to produce energy [4].

There is a variation to attain the process of anaerobic digestion because the waste vary to produce energy as well the operating temperature. So production rate of biogas varies for each case [5].

Anaerobic Digestion is one of the method used to treat food waste. In rural areas the commonly used process is "anaerobic digestion of cow dung" but there are factors that limits the production and technical feasibility to continue the process [6]. Some of the factors affecting anaerobic digestion performance include physical structure of plants, the input to these plants and especially the operating temperature. The input feedstock includes size of particle, content of moisture, content of nutrient and biodegradability [7].

2 Methods

The step by step complete process is defined. Initially the collected kitchen waste is dumped into 4 litre shredder which consist of sharp blades which chopped the waste. The reason is to make process is smoother and less time consuming to generate a biogas. Once the significant amount of waste is gathered in digester tank, the process of anaerobic digestion starts [8]. The process is very precise we have to completely seal the digester. The digester have one inlet and four outlets. These inlet and outlets are connected with PVC pipes and valves. The input valve is connected to waste shredder, a valve placed at the bottom whose purpose is to remove the sludge. Above the sludge removal, fertilizer output valve is attached, this valve is used when level detector indicates that the digester has reached the set level. Meanwhile the input to digester tank is stopped and we reduce the matter through fertilizer outlet and then we can resume the process. This is a rich fertilizer and very good for the growth of plants.

The gas valves placed on top of digester, valve1 is connected to the purification filter while the valve 2 is for emergency if excess amount of gas is produced the emergency valve is connected to desulfurizer (filter) [9]. Valve 1 Connected to the purification filters to remove CO_2, CH_4, H_2S gas. The valve1 is connected to three filters named as desulfurizer, CO_2 scrubber and moisture remover. Desulfurizer contain a sponge of steel wool which removes H_2S, then the gas is proceeded towards the CO_2 scrubber which contains water to remove CO_2 and finally to remove moisture we add moisture remover

* Corresponding author: tanzila@szabist.edu.pk

filter which contain silica gel. After the three stages of filter purified biogas is stored further connected to stove. [10] [11]. The basic process diagram of biogas is shown in figure 1.

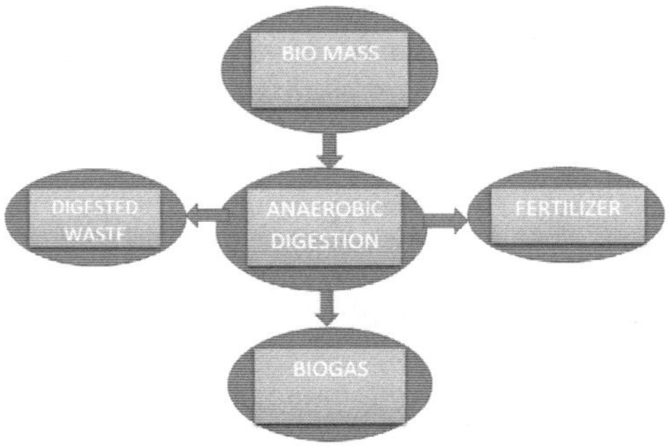

Figure 1. Complete process diagram.

3 Discussion

3.1. Experimental Setup.

The biogas plant is divided into four stages.

1. Waste Shredder
2. Digester Tank
3. Purification
4. Storage

3.2. Parts involved in project.

Following are the parts which are used in our project.

3.2.1. Overall structure of the project.

The structure for the portable biogas plant is of the dimension 6 by 4 feet. It is movable because it has 4 caster wheel joined to it. The biogas plant can be placed anywhere where the sunlight is accessible since it is portable. The dimensions of structure is shown in Table 1. The structure is shown below in figure 2.

Table 1. Structure dimension

S.NO	Parameter	Size
01	Length	6 feet
02	Breadth	4 feet
03	Height	3 feet

3.2.2. Digester dimension.

It is the thick plastic drum as a digester chamber in which waste is added for anaerobic digestion process. The dimensions of digester is shown in Table 2. The digester is shown below in figure 3.

3.2.3. Waste shredder

This drum is connected to a shredder which has a capacity of 4 liters and it contains two blades (saw tooth) for fine chopping. The dimensions of shredder is shown in Table 3. The shredder is shown below in figure 4.

Figure 2. overall structure of project.

Table 2. Digester dimension

S.NO	Parameter	Size
01	Capacity	208 litre
02	Length	2.91feet
03	Diameter	2 feet

Figure 3. Biogas digester.

Table 3. Shredder dimension

S. No.	Parameters	Size
01	Capacity	3 litre

https://doi.org/10.1051/e3sconf/20185101002

Figure 4. Waste shredder.

3.2.4. Purification filters

Then there are three filters biogas desulfurizer, biogas CO_2 and biogas moisture remover which will be used for purification. The dimensions of filters is shown in Table 4. The filters is shown below in figure 5.

Table 4. Filter dimension

S.NO	Parameters	size
01	Jerry can(biogas CO_2 scrubber)	20 litres
02	biogas desulfurizer	1.2 feet
03	Biogas moisture remover	0.9 feet

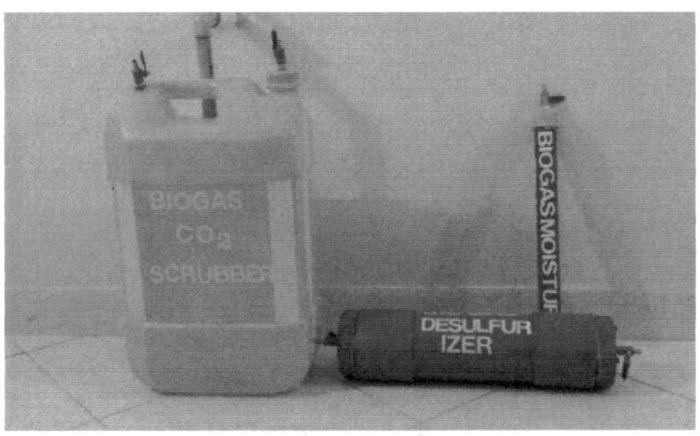

Figure 5. Purification filters.

3.2.5. Biogas storage tube

There is a tire tube is used as storage tank. The storage tube is shown below in figure 6 and tube dimensions is shown in table 5.

Table 5. Storage tube dimension

S. No.	Parameters	Size
01	Tube diameter	16 inch

Figure 6. Storage tube.

3.3. Electronics Part

Following are the electronics components used in the project.

3.3.1. Digester level detector

Digester level detector consist of transistor, timer IC and the buzzer. The level detector is shown below in figure 7.

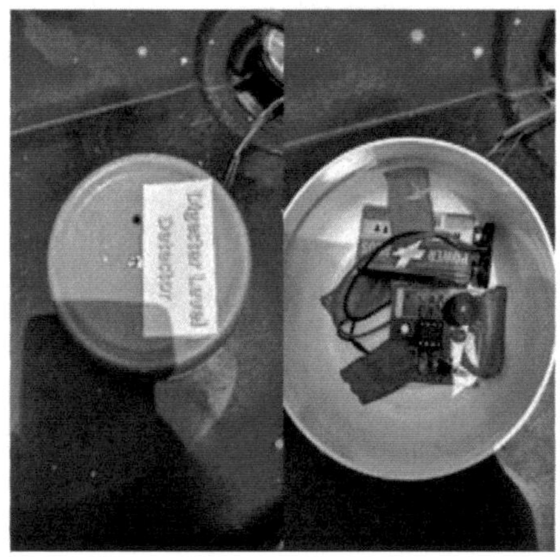

Figure 7. Digester level detector.

3.3.2. Gas leakage detector

Gas leakage detector consist of Arduino, gas sensor and indication circuit. The leakage detector is shown below in figure 8.

3.4. Working principle

Organic matter prevail in form of biological material which is dead or living organisms. These organic substances usually compose of carbon and oxygen, sulphur hydrogen, nitrogen they combine to form many different such as proteins, lipids and carbohydrates. These microorganism break down by nature through digestion in smaller complex carbon substances. Anaerobic process is the digestion mechanism which

 https://doi.org/10.1051/e3sconf/20185101002

happens without oxygen which generated different gases. By anaerobic mainly methane is produces which burns easily at room temperature and also it is eco-friendly which remove dependency on fossil fuels [12] [13].

Figure 8. Gas leakage detector.

4 Results

For observation and analysis, we observe waste material and gas storage calculation need some basic values to calculate the results of experiments which are calculated below. The waste material is listed in Table 6

Table 6. Waste material

Waste material
Banana peel
Carrots peel
Okra
Apple gourd
Ridge
Fenugreek
Tomatoes
Lettuce
Egg plant
Capsicum
Arwi
Potato

In this analysis in Table 7 we are observing change in diameter of tube in daily basis for checking gas generation. The result is shown in Table 7. A figure of biogas project can be shown in Figure 10.

Table 7. Result of tube diameter.

S.NO	Days	Tube Diameter (Inch)
1	1	5
2	2	8
3	3	9
4	4	12
5	5	14
6	6	16
7	7	19.5
8	8	20
9	9	20
10	10	21
11	11	22
12	12	23

Figure 9, shows the graphical image of change in diameter of tube.

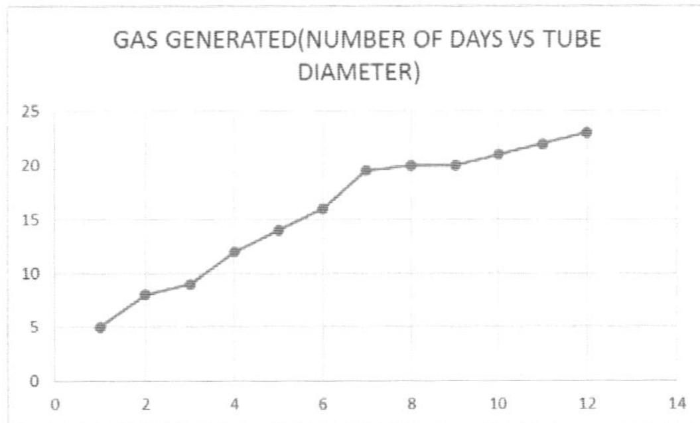

Figure 9. Graphical representation of gas generation.

Figure 10. Biogas complete project.

5 Conclusion

The useful implementation of biogas portable plant for production of biogas by decomposing kitchen waste provide a solution of proper rigid waste management system. Its low price and it works independently and suitable considered parameters which show that it is economic. It is a technology that can be used for processing kitchen waste using a biogas digester .It has suddenly gained importance in the recent time and is a strong competitor in becoming the next renewable source of energy. This plant is more functional in urban areas as more amount of organic waste is produced in urban region because of larger population. Whereas it is value in rural regions because there is shortage of gas supply. In future it could decrease the dependency on CNG and LPG and thus future generations will not be depended on imported fossil fuel [14] [15] .

References

1. D.Vikrant,P.Shekhar,"comparative study of biogas",International Journal of Engineering Science Invention, University of Pune, India,Vol2, 2013.
2. Q. Zhao, E. Leonhardt, C. MacConnell, C. Frear and S. Chen, "Purification Technologies for Biogas Generated by Anaerobic Digestion", CSANR Research Report 2010 – 001
3. Karve of Pune A.D "Compact biogas plant compact low-cost digester from waste starch". www.bioenergylists.org, (2006).
4. T.G.Jantsch, B.Matttiason, "An automated spectropphoyometric system for monitoring buffer capacity in anaerobic digestion processes"Water Research,Vol.38,p.3645- 3650,(2004).
5. D.Gloria Alagcan, S.K.Pratap, M.M.Alagcan, Ginasha, D.R, N.S.Tuivavalagi, S.K. Garg, "Use of Compact Biogas Plant for Biogas Production Utilizing Waste Food Materials, Fruits, and Vegetable Peelings of High Calorific Contents", International Journal of Engineering, ijesm Science and Metallurgy, Vol.2, 2012
6. A.C.Wilkie, "Biomethane from Biomass, Biowaste, and Biofuel", ASM Press,Washington DC,p.195-205,2008
7. S.Chamarthi, S.C.Purnanand, V.Bhale, N.S.Reddy, C.C.Mouli, "Laboratory Scale Experiments for Biogas Production using Gas Chromatography Analysis", IOSR Journal of Engineering, Vol. 3,2013
8. V.Okudoh, "The potential of cassava biomass and applicable technologies for sustainable biogas production in South Africa", Renewable and Sustainable Energy, Vol 39, p.1035–1052,2014
9. P.Gauri Minde, S.Sandip, Magdum and V. Kalyanraman, "Biogas as a Sustainable Alternative for Current Energy Need of India", Journal of Sustainable Energy & Environment, Vol 4, p.121-132, 2013.
10. D.Gloria Alagcan, S.K.Pratap, M.M.Alagcan, Ginasha, D.R, N.S.Tuivavalagi, S.K. Garg, "Use of Compact Biogas Plant for Biogas Production Utilizing Waste Food Materials, Fruits, and Vegetable Peelings of High Calorific Contents", International Journal of Engineering, ijesm Science and Metallurgy, Vol.2, 2012.
11. H. Igoni, M. F. N. Abowei, M. J. Ayotamuno and C. L. Eze, "Effect of Total Solids Concentration of Municipal Solid Waste on the Biogas Produced In an Anaerobic Continuous Digester", 2008.
12. A.Ali, L.Rundong, "Predictive Modeling of Biogas Production from Anaerobic Digestion of Mixed Kitchen Waste at Mesophilic Temperature", International Journal of Waste Resources,2016
13. D.Gloria Alagcan, S.K.Pratap, M.M.Alagcan, Ginasha, D.R, N.S.Tuivavalagi, S.K. Garg, "Use of Compact Biogas Plant for Biogas Production Utilizing Waste Food Materials, Fruits, and Vegetable Peelings of High Calorific Contents", International Journal of Engineering, ijesm Science and Metallurgy, Vol.2, 2012
14. D.Gloria Alagcan, S.K.Pratap, M.M.Alagcan, Ginasha, D.R, N.S.Tuivavalagi, S.K. Garg, "Use of Compact Biogas Plant for Biogas Production Utilizing Waste Food Materials, Fruits, and Vegetable Peelings of High Calorific Contents", International Journal of Engineering, ijesm Science and Metallurgy, Vol.2, 2012
15. D.Gloria Alagcan, S.K.Pratap, M.M.Alagcan, Ginasha, D.R, N.S.Tuivavalagi, S.K. Garg, "Use of Compact Biogas Plant for Biogas Production Utilizing Waste Food Materials, Fruits, and Vegetable Peelings of High Calorific Contents", International Journal of Engineering, ijesm Science and Metallurgy, Vol.2, 2012

Characterization of Laciana Valley District mine water as geothermal resource

Alejandro Matas[1,2,*], Francisco García-Carro[2] and Jorge Loredo[1]

[1] Department of Mining Methods and Prospection, University of Oviedo, C/Independencia, 13, 33004 Oviedo, Asturias, Spain.
[2] I+D+i Department at Magna Dea, S.L. C/Posada Herrera 2,2 33002 Oviedo, Asturias, Spain.

Abstract. Laciana Valley District is a coal mining region located in northern Spain. This region counts with several mining facilities, among them: eight underground mines (six mountain mines and two mine shafts) and three open pit mines. Nowadays, all the mining facilities have been closed down and are flooded. The water found inside them could be used as a geothermal resource due to its thermal properties and the proximity to population. The aim of the study is to analyze the water of the facilities located in Laciana Valley and determine its potential to be used as a geothermal resource for a district heating system. In order to achieve this goal an extensive field work has been performed, nine different mine water discharges have been chosen and several water characteristics have been selected for analysis. The parameters measured have been pH, conductivity, hardness, temperature, turbidity and alkalinity. The results have been evaluated in order to determine the fluctuations of the physico-chemical parameters throughout a hydrological year and the mining facilities have been compared between each other regarding their mine water quality. The analysis of all the information gathered in the study shows a noteworthy thermal potential in the water of the abandoned mines of the region.

1 Introduction

Flooded underground mines provide an outstanding opportunity to extract low-grade geothermal energy. In the last ten years numerous research and commercial initiatives have been undertaken in Europe to develop abandoned coal mining fields into low-temperature resources [1]. This paper is focused on Laciana Valley District´s coal mine facilities, located in northern Spain. Nowadays, all the mining facilities have fallen into disuse and are flooded. Coal mining began in the beginning of the 20th century and continued as the primary industry of the town until 2008 when the global economic crisis took its toll. With the main industry removed, the region entered a period of economic and social readjustment and had to strive to develop new industries. The use of mine water as a renewable heating source could contribute to the sustainable regeneration of former mining areas, providing both jobs and secure, low carbon heating energy [2].

The application of mine water for heating and cooling applications has been well known for quite some time, but still not many installations have been documented at a world scale, as of 2013 only 20 were documented [3].

According to [4] there are some advantages of flooded mines as compared to natural aquifers:

- excellent heat exchange surface (area) from rock to water caused by mining,
- the geological situation is usually well known and documented,
- great volumes (x 100,000 m^3) of impounded mine water (partly > 20°C),
- high permeability and productive discharge of water (dekalitre/s),
- shafts and tunnels for the installation of pumps and heat pump systems exist and are accessible without bigger efforts,
- suitable locations for the installation of pumps and for re-infiltration of cooled water.

The first mayor abandoned coal mine water power station opened at Heerlen in the Netherlands in 2008. It currently uses 32 °C water extracted from underground coal workings through boreholes in the Dutch coal mining heartland. This scheme is now used to heat three hundred homes and businesses, with an estimated 55% reduction in CO2 emission compared to conventional water heating systems [1].

Previously to the achievement of the Mine water Project in Europe there has been several mine water heat recovery projects carried out as stated by [3]. In the late 1980s the mine water of a flooded underground coal mine was used as a geothermal resource at Springill, Canada. Mine water at 18°C supplied to heat pumps used to heat and cool several buildings including manufacturing facilities and a restaurant [5]. In Germany there have been numerous underground mines whose water have been used to heat public buildings such as nursing homes, high schools or swimming pools [4]. These mines are

* Corresponding author: jloredo@uniovi.es

located in the Rhenish Massif (Essen and Sachsen) where a total number of five projects have been implemented between the late 1980s and 2007. It has been estimated that 3000 MWt heat energy is available in the waters of flooded coalfields of Europe [6]. Nevertheless, the use of mine water as a renewable heating source has been also appealing out of Europe, there are notorious facilities in USA and Canada such as a flooded lead mine in Park Hills, Missouri, where water is extracted from workings between 11 and 133 m below ground level and has been used to heat and cool a Municipal Building (area 753 m2) since 1995 [7].

More recently, there has been a powerful impetus in Asturias, Northern Spain, provided by the private company HUNOSA to focus on heat recovery in its mines. According to [8], mine water reservoirs in Central Asturias could reach a potential of 40 million m3 per year for water supply, and an energy supply capacity close to 260,000 thermal MWh per year. Mine water from a mining reservoir (Barredo-Figaredo) has been already used as geothermal resource to heat and cool some University buildings and there are ongoing projects to extend these applications to other potential future users in the area [9].

UK has been another major promoter of this kind of energy, the Scottish Parliament commissioned a report on the energy and environmental benefits of using water–water heat pumps, with special reference to the waters of abandoned coal mines [8]. In Shettleston, Scotland mine water at 12°C, from 100 m deep borehole drilled into flooded underground coal mine workings, is used for space heating and pre-heating of domestic hot water for 16 residential properties (1600 m2 total). In Lumphinnans, also Scotland, mine water at 15°C supplied to heat pumps of rated capacity 65 kW, producing hot water at 55°C. Used for space heating and pre-heating of domestic hot water for 16 residential properties (1600 m2 total) [7]. In South Wales region approximately three thousand litres of water are being drained or pumped from old mine workings per second. Using discharge and temperature data it is estimates that this is 72MWof potential heating energy [10].

These examples provide good proof of the potential of underground mine waters as a geothermal resource. The heat in mine water is a valuable resource that is currently being wasted when it is discharged to a receiving stream. Acid mine drainage (AMD) causes environmental pollution that affects many countries having historic or current mining industries [11]. The effective harnessing of energy from mine water, either from naturally overflowing mines or pumped shafts/boreholes, can turn an environmental liability (mine water) into an environmental asset (a clean energy source) [7]. The extraction of energy from them, which is immediately returned to the mine, produces no combustion gases and leaves no chemical residue on the surface. This energy source, when in routine use, produces no harm to the local environment [5].

2 Area of study

This study focusses on Laciana Valley District abandoned mines, located in Northern Spain, close to the city of Villablino (Fig. 1).

The geology of Laciana region consists of a thick (up to 3.000 m) sedimentary sequence of Estephanian age. They crop out along a narrow band of 4 km wide and up to 25 km long. This unit is known as "The Villablino basin" and is supported by Precambrian and Palaeozoic materials. Main lithofacies include mudstones, siltstones, sandstones (litharenites), coal seams and minor quantities of limestones and conglomerates. Coal seams thickness varies from 0,40 to 2 meters, reaching 3 meters in extraordinary cases (Paulina mountain mine) [12].

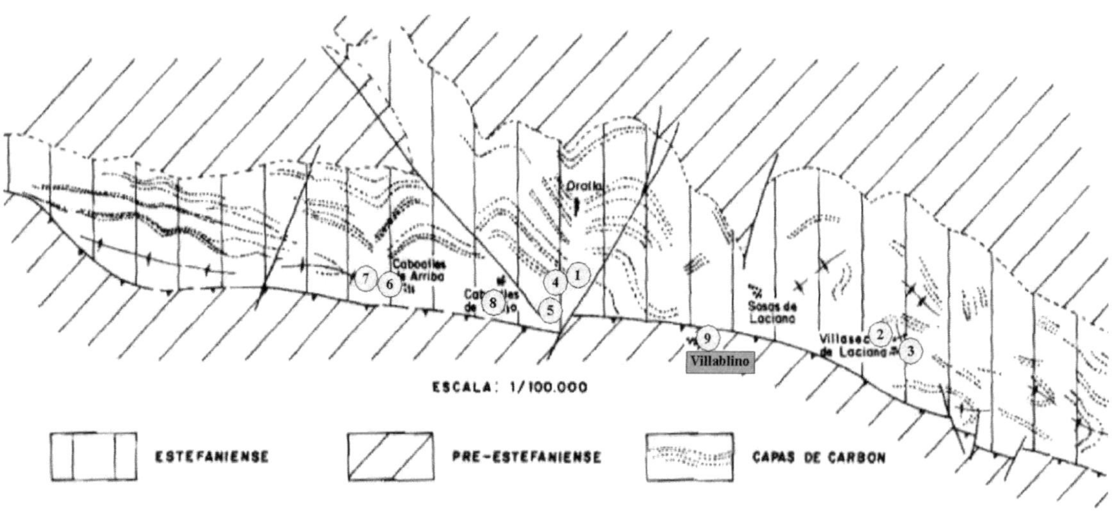

Figure 1. Geological scheme of Laciana mining area and sampling points location [12].

There are eleven different mining facilities (eight underground mines and three open pit mines). Nine naturally overflowing points have been selected to analyze the mine water characteristics in order to research their value as low enthalpy geothermal resource:

1. Calderón mountain mine
2. Lumajo mountain mine
3. Carrasconte mountain mine
4. Calderón transformers
5. Bolsada Caboalles de Abajo mountain mine
6. Bolsada Caboalles de Arriba mountain mine
7. Paulina mountain mine
8. María mine shaft
9. Transversal Villablino mountain mine

3 Materials and methods

Four sampling campaigns were performed, covering a full hydrological year (in April–July-November-January 2017-2018). Nine sampling points were identified and detailed in point 2. All of the sample points were selected based on criteria such as accessibility, solicitation and location in the vicinity of population. Before a sample was collected, the bottle was rinsed three times with the water to be sampled. During the site inspection all sampling points were photographed and documented and field measurements of pH, conductivity, hardness, temperature and turbidity were recorded. Temperature and pH were measured with a Hanna instrument (Waterproof pH meter HI9024C), turbidity was obtained with a Hanna´s turbidity meter HI 93703, conductivity was collected with Hanna´s conductivity meter HI9033, hardness was measured through a JBL total hardness test and alkalinity was obtained through a JBL alkalinity test. Equipment was calibrated before each day's fieldwork and all water samples were refrigerated as soon as possible after collection.

4 Results and discussion

4.1 Physicochemical characteristics

Table 1 shows temperature and geochemical data for all mine waters that were included in this analysis. The vast majority of mine water flows treated has circum-neutral pH (overall mean: 7.0; range: 6.36–7.55). Most of the sites display small temperature variations during the monitoring period; however variation between the study sites is bigger than 5 °C. The lowest temperatures are recorded at Bolsada Caboalles de Arriba mountain mine (9.8 °C) while the warmest temperature (15.5 °C) was recorded at Calderón transformers. Warm water was also found in Calderón mountain mine (15.2°C), María mine shaft (14.1°C) and Transversal Villablino mountain mine (14,7°C). Values of turbidity differ substantially from a study site to another. A high turbidity can be found in Carrasconte (83 NTU) mountain mine, whereas in Calderón mountain mine or Transversal Villablino mountain mine waters have low turbidity (2,29 and 2,11 respectively). The conductivity variation is reasonably stable throughout the year. The low values (< 2000 µS/cm) of the most of the sampling sites are in line with the conductivity found in shallow fresh ground waters. Mine waters from greater depth tend to have higher conductivities due to longer rock-water interaction, greater potential influence of saline waters, and inflows of strata water with higher conductivity [13]. All the waters evaluated can be classified as very hard according with their hardness (400-1000 µS/cm). Lumajo mountain mine and Bolsada Caboalles de Arriba mountain mine are the samples sites with softest waters, nevertheless their hardness is over 250 mg/l $CaCO_3$. Most of the samples show alkalinities lower than those found in mine water discharges of northern Spain [14]. However, Carrasconte, Calderón transformers and María mine shaft alkalinity is higher than the others and similar to the ones found in Asturias [14].

4.2 Laciana Valley District´s mine water as a geothermal resource

Compared to known systems using mine water as a heat source the conditions in the monitored mines in Laciana region seem to be positive. Watzlaf and Ackman [15] list several locations where mine water is used in geothermal systems at temperatures comparable or lower than those reported in the current paper. In Springhill, Nova Scotia, Canada, water is used at temperature of 18 °C. In Park Hills, Missouri, United States, heat pumps works with mine water at 14°C. In Shettleston, Scotland, United Kingdom, mine water at 12°C is used. In Lumphinnans, Scotland, United Kingdom, mine water used for heating has 14.5°C. In Markham Colliery, Derbyshire, Alkane Energy is using geothermal heat from mine water at 15°C [16]. Economic assessments have been performed in several of the locations cited. For instance, the payback period in Park Hills was calculated to be about 4.6 years [15] and the payback period for the additional capital cost in Springhill was estimated to be less than one year [5].

In order to determine the scaling and corrosion potential of Laciana Valley District mine water in a heat pump system, the Langelier Saturation Index (LSI) and the Ryznar Stability Index (RSI) are going to be calculated. These indices are commonly used in the water treatment industry to evaluate the nature of a water source [17] and their interpretation is shown in table 2. Total dissolved solids (TDS), pH, temperature, Calcium hardness and M alkalinity are necessary to calculate the Langelier Saturation Index. All the necessary parameters have been obtained from the analysis carried out but TDS which has been calculated using two different methods.

On one hand, TDS includes all solids that are dissolved in water, hence, its value is very similar to the sum of the quantities of the constituents of the water analyzed separately [18]. Hardness is represented mostly by Ca^{2+} and Mg^{2+}, while alkalinity is represented mostly by HCO_3^- [17]. Therefore TDS must be at least as high as the addition of hardness and alkalinity in ppm or

Table 1. Physicochemical data for the sampling points identified.

Sample	pH	Temperature (ºC)	Turbidity (NTU)	Conductivity (µS/cm)	Hardness (mg/l CaCO3)	Alkalinity (ppm CaCO3)	TDS (ppm CaCO3)
Calderón mountain mine							
April 17	7,1	14,1	1,03	1343	836,6	327	1163,6
July 17	6,73	15,2	2,29	1602	836,6	305,2	1141,8
November 17	7,55	15,1	1,61	1281	729,8	327	1056,8
January 18	7,38	13,6	0,65	1349	890	370,6	1260,6
Mean	7,19	14,50	1,40	1393,75	823,25	332,45	1155,7
Lumajo mountain mine							
April 17	6,98	12,5	15,85	649	462,8	327	789,8
July 17	6,67	12,9	14,22	540	284,8	261,6	546,4
November 17	7,17	13,1	5	408	267	196,2	463,2
January 18	7,33	13,1	6,82	515	231,4	239,8	471,2
Mean	7,04	12,90	10,47	528,00	311,50	256,15	567,65
Carrasconte mountain mine							
April 17	6,45	12,5	83	1840	1121,4	610,4	1731,8
July 17	6,5	12,7	47,51	1767	854,4	545	1399,4
November 17	6,71	12,5	64	1358	1068	588,6	1656,6
January 18	6,71	12,3	13,4	1690	907,8	566,8	1474,6
Mean	6,59	12,50	51,98	1663,75	987,90	577,70	1565,6
Calderón transformers							
April 17	7,03	14,3	10,79	1541	854,4	457,8	1312,2
July 17	6,87	15,5	10,48	2340	996,8	501,4	1498,2
November 17	7,08	15,3	16,92	2070	1192,6	545	1737,6
January 18	7,17	13,5	5,57	1787	783,2	457,8	1241
Mean	7,04	14,65	10,94	1934,50	956,75	490,50	1447,25
Bolsada Caboalles de Abajo mountain mine							
April 17	6,81	12,3	6,92	923	534	305,2	839,2
July 17	6,65	12,4	5,83	1123	534	305,2	839,2
November 17	7,05	12,3	2,16	828	605,2	261,6	866,8
January 18	6,86	11,9	8,46	1306	676,4	327	1003,4
Mean	6,84	12,23	5,84	1045,00	587,40	299,75	887,15
Bolsada Caboalles de Arriba mountain mine							
April 17	7,13	9,8	0,48	233	249,2	130,8	380
July 17	6,85	9,8	30,53	397	231,4	87,2	318,6
November 17	7,23	10	2,33	312	267	130,8	397,8
January 18	7	10,2	7,48	182	106,8	65,4	172,2
Mean	7,05	9,95	10,21	281,00	213,60	103,55	317,15
Paulina mountain mine							
April 17	6,59	10,6	19,17	527	302,6	261,6	564,2
July 17	6,36	10,3	6,21	458	231,4	152,6	384
November 17	6,61	9,6	0,5	365	320,4	196,2	516,6
January 18	6,86	8,7	21,73	247	142,4	130,8	273,2
Mean	6,61	9,80	11,90	399,25	249,20	185,30	434,5
María mine shaft							
April 17	6,9	14	8,89	806	427,2	457,8	885

July 17	6,61	14,1	12,68	1000	480,6	436	916,6
November 17	6,81	13,7	12,01	762	462,8	501,4	964,2
January 18	6,89	13,7	8,69	992	462,8	523,2	986
Mean	6,80	13,88	10,57	890,00	458,35	479,60	937,95
Transversal Villablino mountain mine							
April 17	6,71	14,5	2,11	765	373,8	305,2	679
July 17	6,5	14,6	1,32	878	391,6	348,8	740,4
November 17	7,29	14,7	0,14	703	445	370,6	815,6
January 18	7,3	14,7	0,05	845	409,4	348,8	758,2
Mean	6,95	14,63	0,91	797,75	404,95	343,35	748,3

milligram per liter. On the other hand there is a relation between TDS expressed in milligram per liter and electrical conductivity in microsiemens per centimeter at 25°C. The correlation factor varies between 0.55 and 0.80 and is higher with bigger values of TDS and conductivity [19]. An estimation with the factor 0.64 has been made, as proposed by [20].

LSI and RZI have been calculated resulting very similar for both approximations of the TDS. The results displayed in tables 1 and 3 consider TDS as the addition of alkalinity and hardness.

As shown in table 3, LSI of Laciana Valley District mine waters is mostly slightly corrosive or balanced. Nevertheless some locations such as Bolsada Caboalles de Arriba y Paulina mountain mine have nearly serious corrosive waters and Calderón mountain mine and Calderón transformers have balance or slightly scaling waters

5 Conclusion

Laciana Valley District is a mining region, located in northern Spain, which counts with several abandoned mining facilities. The mine water of nine different effluent points has been analyzed in this paper. Very stable mine water temperatures throughout the year and relatively high value of water temperature at the surface discharge points show that these waters have great potential for geothermal energy generation. Several geothermal systems from around the globe exploit mine water at temperatures comparable or lower than those reported in this paper. The different parameters analyzed (temperature, pH, conductivity, hardness, turbidity and alkalinity) show relatively low variations during the year but present important fluctuations from one sample site to another. Calderón mountain mine and Calderón transformers have been proved to be the best options for heat recovery system installations, followed by Transversal Villablino mountain mine and María mine shaft. The relative stability of the studied parameters suggests that these waters represent a dependable thermal resource that, given current levels of demand, can be sustainably exploited for long-term geothermal use. However, it must be cautioned that some of these waters can cause corrosion problems in pipes and heat exchangers and some actions to prevent them should be applied. The results of this study are a sound base for further evaluation of Laciana Valley District mine water as geothermal resource. The evaluation of profitable water treatments and an economic assessment of the facilities might be fruitful topics to develop.

Table 2. Interpretation of the Langelier Saturation Index and Ryznar Stability Index [21].

LSI Index Value	Indication	RSI Index Value	Indication
2,00	Scale forming but non corrosive	4.00-5.00	Heavy scale
0,50	Slightly scale forming and corrosive	5.00-6.00	Light scale
0,02	Balanced but pitting corrosion possible	6.00-7.00	Little scale or corrosion
-0,5	Slightly corrosive but non-scale forming	7.00-7.50	Corrosion significant
-2,00	Serious corrosion	7.50-9.00	Heavy corrosion
		> 9.00	Corrosion intolerable

Table 3. LSI and RSI of Laciana region mine water.

Sample	Langelier Saturation Index (LSI)		Ryznar Stability Index (RSI)	
Calderón mountain mine	0,43	(Slightly scale forming and corrosive)	6,33	(Little scale or corrosion)
Lumajo mountain mine	-0,26	(Balanced but pitting corrosion possible)	7,56	(Heavy corrosion)

Carrasconte mountain mine	0,10	(Slightly scale forming and corrosive)	6,40	(Little scale or corrosion)
Calderón transformers	0,50	(Scale forming but non corrosive)	6,03	(Little scale or corrosion)
Bolsada Caboalles de Abajo mountain mine	-0,14	(Balanced but pitting corrosion possible)	7,13	(Corrosion significant)
Bolsada Caboalles de Arriba mountain mine	-0,84	(Slightly corrosive but non-scaleforming)	8,73	(Heavy corrosion)
Paulina mountain mine	-0,98	(Slightly corrosive but non-scaleforming)	8,57	(Heavy corrosion)
María mine shaft	-0,06	(Balanced but pitting corrosion possible)	6,92	(Little scale or corrosion)
Transversal Villablino mountain mine	-0,08	(Balanced but pitting corrosion possible)	7,12	(Corrosion significant)

References

1. R. Verhoeven, E. Willems, V. Harcouët-Menou, E. De Boever, L. Hiddes, P.O. t. Veld, E. Demollin, Energy Procedia **46** 58–67 (2014).
2. G. Farr, S. Sadasivam, Manju, I.A. Watson, H.R. Thomas, D. Tucker, Int. J. Coal Geol. **164** 92–103 (2016).
3. M. Preene, P.L. Younger, Min. Technol. **123** 107–118 (2014).
4. G. Wieber, S. Pohl, Int. Mine Water Assoc. Congr. 113–116 (2008).
5. A. Jessop, Proceedings **17** 463–468 (1995).
6. R. Díez, M. Díaz-Aguado, Energies **7** 4241–4260 (2014).
7. D. Banks, H. Skarphagen, R. Wiltshire, C. Jessop, Geol. Soc. London, Spec. Publ. **236** 499–513 (2004).
8. S. Jardón, A. Ordóñez, R. Álvarez, P. Cienfuegos, J. Loredo, Mine Water Environ. **32** 139–151 (2013).
9. J. Loredo, A. Ordóñez, S. Jardón, R. Álvarez, Imwa 2011 177–182 (2011).
10. D.J. Thomas, Fuel **189** 445 (2017).
11. D.B. Johnson, K.B. Hallberg, Sci. Total Environ. **338** 3–14 (2005).
12. IGME, Mapa Geológico de España E. 1:50.000 Villablino, Spain, 1982.
13. M.T. Bailey, C.J. Gandy, I.A. Watson, L.M. Wyatt, A.P. Jarvis, Int. J. Coal Geol. **164** 77–84 (2016).
14. C. Loredo, A. Ordóñez, E. Garcia-Ordiales, R. Álvarez, N. Roqueñi, P. Cienfuegos, A. Peña, N.M. Burnside, Sci. Total Environ. **576** 59–69 (2017).
15. G. Watzlaf, T. Ackman, Mine Water Environ. **25** 1–14 (2006).
16. A.P. Athresh, A. Al-Habaibeh, K. Parker, Int. J. Coal Geol. **164** 69–76 (2016).
17. K. Rafferty, Geo-Heat Center, Klamath Falls, OR 11–16 (2000).
18. C.S. Howard, Ind. Eng. Chem. Anal. Ed. **5** 4–6 (1933).
19. E.A. Atekwana, E.A. Atekwana, R.S. Rowe, D.D. Werkema, F.D. Legall, J. Appl. Geophys. **56** 281–294 (2004).
20. J.U. Iyasele, D.J. Idiata, D. J, Int. J. Eng. Res. Rev. **3** 40–48 (2015).
21. Carrier, *Handbook of Air Conditioning System Design* (1965)

Coastal protection of the Romanian nearshore throughout hybrid wave and offshore wind farms

Alina Raileanu, Florin Onea* and Liliana Rusu

Department of Mechanical Engineering, ''Dunarea de Jos'' University of Galati, Domneasca" Street, 47, Galati 800008, Romania

Abstract. The objective of the present work is to estimate the influence of several hybrid wind and wave farm configurations on the wave conditions reported in the vicinity of the Saint George coastal area, in the Romanian nearshore of the Black Sea. Based on the wave data coming from a climatological database (ERA20C) and also on in situ measurements, it was possible to identify the most relevant wave patterns, which will be further considered for assessment. The numerical simulations were carried out with the SWAN (Simulating Waves Nearshore) wave model, which may provide a comprehensive picture of the wave transformation in the presence of the marine farms. Although the impact of the wind farm is not visible from the spatial maps, from the analysis of the values corresponding to the reference points, it was noticed that a maximum variation of 2% may occur for several wave parameters.

1 Introduction

Romania is defined by numerous opportunities in terms of the renewable energy sources, from which it can be mentioned the hydropower and wind energy. With a generated capacity of 1894 MWh (reported to 2015), the hydropower sector has a share of 26% from the total electricity production, from which a significant percentage is obtained throughout one of the largest project from Europe (the Iron Gate I power station) [1]. As for the wind energy (reported to 2014), it was estimated that almost 75 of the wind farms operates in Romania, most of the projects being implemented in the Dobrogea Plateau (78 %), which is an geographical area located close to the Black Sea [2].

There is a close connection between wind and waves, and during the recent years it was highlighted that the western part of the Black Sea seems to be defined by consistent marine conditions which can be extracted efficiently throughout wind and wave farms, or eventually throughout hybrid wind-wave projects [3-5]. In the case of the Romanian nearshore, the combination between the wave action and the reduced sediment flux due to the presence of the dams on the Danube River has accentuated the local coastal erosion [6]. In order to tackle this issue, a possible solution will be to consider the use of some Wave Energy Converter (WECs) which may be deployed in a wave farm capable to attenuate the wave power which is propagated throughout the surf area. Similar approaches were proposed for some other coastal environments, such as in the case of the Portuguese nearshore or for the western sector of the Mediterranean Sea [7-9]. If we consider the progress reported by the wave industry and the attractiveness of the wave power, it is expected that in the near future this source of energy will become more competitive, being possible to develop wave farms in enclosed basins (such as the Black Sea) defined by less energetic conditions than the ones reported in the ocean environments. Nevertheless, if we discuss about the western part of the Black Sea it is more feasible to consider that an offshore wind farm has better chances to be developed first, being possible in this way to support the development of a pilot wave project.

In this context, the objective of the present work will be simulate the coastal impact of different hybrid wind and wave farm configuration on a target area located in the vicinity of the Danube Delta, which is a protected area included in the UNESCO heritage.

2 Methods and materials

The target area considered for simulation is located on the western part of the Black Sea, more precisely close to the Saint George sector (Romania) which is in the vicinity of the Danube Delta. Figure 1 illustrates the computational domain considered for simulation, and also the proposed case studies (denoted with CS). The numerical simulations will be carried out by using the SWAN (Simulating Waves Nearshore) wave model which is considered to be a third-generation model capable to provide a realistic estimation of the wave parameters from the coastal environments [10, 11]. In the background is presented the bathymetry, where a maximum depth of 50 m may be observed in the offshore region. The computational domain is rectangular, being defined by a length of 14 km in the X-direction, while a 20 km is reported along the Y-axis. In Figure 1a, is presented the offshore wind farm considered for investigation which includes 140 wind turbines, similar to the Greater Gabbard project from UK which is one of the

* Corresponding author: florin.onea@ugal.ro

largest wind farm from this area. The total area covered by the farm is around 147 km^2, while the distance between the piles was set to 650 m, similar to the original project configuration [12]. More details regarding the SWAN setup and the computational domain may be found in Table 1.

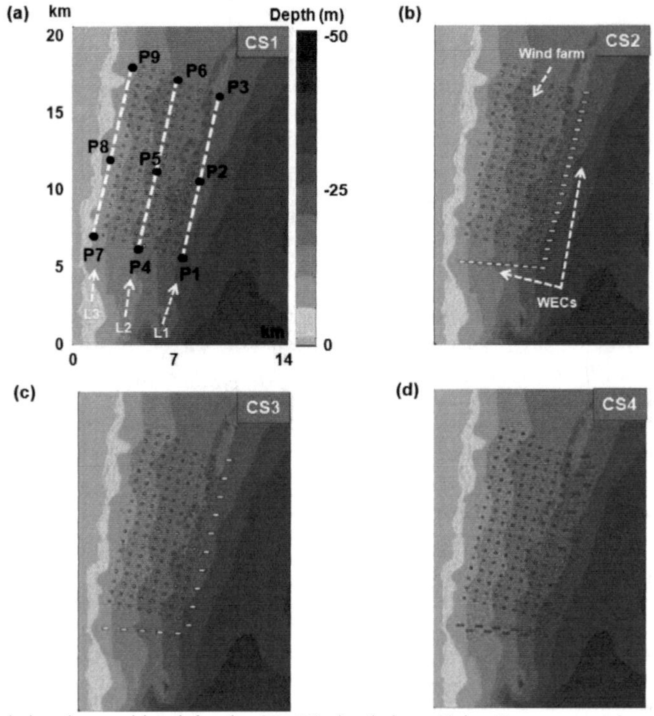

Figure 1. The computational domain considered for the SWAN simulations (Saint George coastal area). In the background, the bathymetry is represented while in the foreground the references lines/points are indicated, as also the selected case studies, where: a) CS1; b) CS2; c) CS3; d) CS4.

Table 1. Characteristics of the SWAN computational domain and description of the physical processes activated.

Computational domain setup		Input/process	
Coordinates: Cartesians	No points in Y-direction: 406	Waves: X	Interactions: Quad; Triads
Δ x (m): 50	No of frequencies: 34	Wind: X	Whitecapp.: X
Δ y (m): 50	No of directions: 35	Current: X	Refraction, Diffraction: X
Δ θ (°): 5	Mode: Stationary	Gen:	Nearshore processes:
No points in X-direction: 355		Westh.	Friction; Breaking; Set up

The designed offshore wind farm will represent the base, around which will be developed several case studies which include wave converters, as it can be observed from Figure 1b where a single line of WECs was added. For the current work was considered useful to use the technical characteristics of the Wave Dragon device, which is one of the largest system on the market being defined by a total length of 300 m [13]. For this case, the distance between the devices was considered equal to the length of this device, respectively 300 m. For the case study CS3, a similar setup was considered, the difference is that in this case the space between the WECs is double (600 m), which means that a much smaller number of devices were considered (15 WECs). The last case study, denoted with CS3, is based on the CS2 scenario, with the mention that in this case a second line of WECs was added in front of the existing devices. All the WECs are located in front of the wind farm in order to provide a sheltering area for the wind turbines and for the beach area.

More details regarding these case studies are presented in Table 2, where are also mentioned the absorption property of the wind and wave devices, which were included in the SWAN simulation as obstacles [13, 14].

Table 2. Description of the case studies considered for investigation in the SWAN simulations.

Case studies		Absorption scenarios	
CS1	Offshore wind farm – 140 piles; Distance between piles – 650 m.		
		Turbine piles	WECs
CS2	Wind farm+WECs deployed on a single line; Distance between the systems – 300 m (the total length of a Wave Dragon).		
CS3	Wind farm+WECs deployed on a single line; Distance between the systems – 600 m (double the length of a Wave Dragon).	TRANS=0.72	TRANS=0.68
CS4	Wind farm+WECs deployed on two lines; Distance between the systems – 600 m.	REFL=0.2	REFL=0.2

In addition to the case studies configuration, another important step is related to the correct assessment of the wave conditions from the targeted area in order to initiate the SWAN model. Two datasets will be considered for investigation, from which the first one is related to the ERA20C database which is a reanalysis product component of the European Reanalysis of Global Climate Observations project [15]. This data cover the interval from January 1900 to December 2010, while for this study the time series of the significant wave height (*Hs*) and the wave period (*Tm*) were processed for the entire interval (111 years of data) considering a references point located in the vicinity of the target area (45°N/30°E). As for the wave direction, were considered the in situ measurements registered at the Gloria drilling platform (44°31'N/29°34'E), covering the interval from January 2003 to December 2009.

Figure 2 illustrates the monthly distribution of the wave parameters from the vicinity of the Saint George sector, a particular attention being given to the extreme values which are more relevant for the coastal stability. In terms of the *Hs* parameter, we can easily identify the differences between the summer and winter time (October-March), where the last one reveal more energetic values. A maximum of 4.76 m may expect in December, compared to the summer values which during the interval May-August does not exceed 2.3 m. Other important winter months are January and February, during which the *Hs* heights of 4.34 m and 4.49 m may be reached. In Figure 2b is presented the distribution of the waves higher than 1.5 m limit, which is considered to be a representative threshold for the maritime activities being linked to the operational limit of the workboats [16].

In this case the best results are reported during the summer time when a maximum of 1.45% may occur in April, while on opposite side we found January and December with 10.3% and 8.6%, respectively. The wave period varies between 5.8 s and 8.8 s, while in terms of the wave direction an average value of 137.6° (south-east) may be considered representative for the entire year, with the mention that during September may be encountered waves associated with the 102.7° value.

From the analysis of these values was considered interesting to run the SWAN simulation with a sea state associated to the following wave parameters: *Hs* – 4.76 m; *Tm* – 8.8 s; *Dir* – 137.6°.

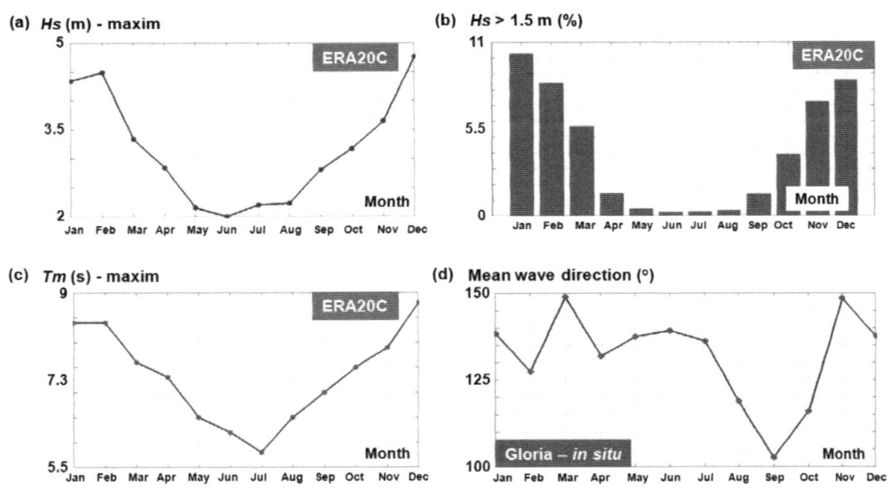

Figure 2. Monthly wave statistics in the vicinity of the target area (Saint George) as resulting from processing the ERA20C database and the Gloria in situ measurements, where: a) *Hs* parameter – maximum values; b) *Hs* parameter – percentage exceeding 1.5 m; c) *Tm* parameter – maximum values; d) Wave direction – average values.

3 Results

Figure 3illustrates the spatial distribution of the *Hs* values in the absence of the farm and for the case study CS1 (wind farm), where was also included the evolution of this parameter along the considered reference lines. In Figure 3a was also included the wave vectors in order to notice the direction of the waves (south-east) used in the simulations, but in the next figures this field will be removed in order highlight more clearly the influence of the wind and wave devices. Several wave fields occur in the target area, their variation being related to the orientation of the water depth isolines being indicated that in the vicinity of the coastline a wave field of 2.5 m may be encountered. From the distribution of the *Hs* heights along the reference lines can be observed the transition of the waves from the offshore area to the shallow water (line 3) where the dissipative effects become more important. The impact of the wind farm seems to be minimal, being noticed no variation in the spatial distribution or along the selected lines.

A complete description of the wave parameters is presented in Table 3 (no farm situation), where with the *Vbot* (m/s) is indicated the bottom velocity, where a zero value represent a truly deep water [17].

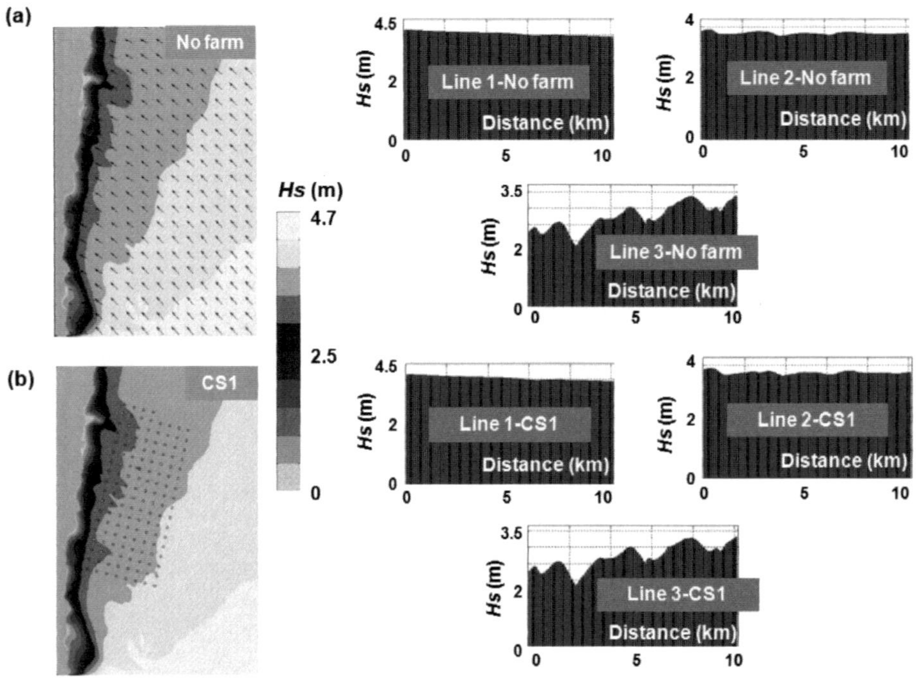

Figure 3. Evaluation in the geographical space of the wave distribution, where: a) No farm; b) CS1 – offshore wind farm similar to the Greater Gabbard project (140 turbines).

Table 3. The values of the wave parameters corresponding to the references points P1-P9. No wind or wave farm was considered for this evaluation, which is representative for the following sea state: *Hs* – 4.76 m; *Tm* – 8.8 s; *Dir* – 137.6°.

Parameter	Reference points								
	P1	P2	P3	P4	P5	P6	P7	P8	P9
Depth (m)	26.06	22.4	19.8	15.3	17.3	16.85	4.814	7.153	8.601
Hs (m)	4.194	4.069	3.932	3.867	3.763	3.772	2.295	2.948	3.374
Tm (s)	7.661	7.734	7.721	7.748	7.765	7.775	8.222	8.076	7.97
Dir (°)	134.7	131.2	129.3	126.4	126.4	128	108.9	112	117
Vbot (m/s)	0.6265	0.7365	0.8144	1.04	0.9	0.9257	1.475	1.464	1.467

https://doi.org/10.1051/e3sconf/20185101004

Figure 4 presents the evolution of the wave fields in the presence of the CS2 scenario. The influence of the WECs is visible throughout the entire wind farm, more important variations being reported along the reference line 1, which is located between the wind and wave farm. Also it seems that on a local scale, the wind turbines seem to generate small shadow areas, aspect which was not noticed in the previous distribution. The *Hs* values reported in the center of the wind farm (line 2) seems to decrease with almost 0.5 m, this variation being constant along this line. For the area located between the wind farm and coastline, it seems that no important variation will occur.

By increasing the spaces between the WECs, the shadow effect of the WECs will significantly decrease, as it can be observed from Figure 5 (CS3 study case). Nevertheless the influence of the wave farm is still visible in the lower part of the wind park (facing the coastline), being also observed along the reference line 1.

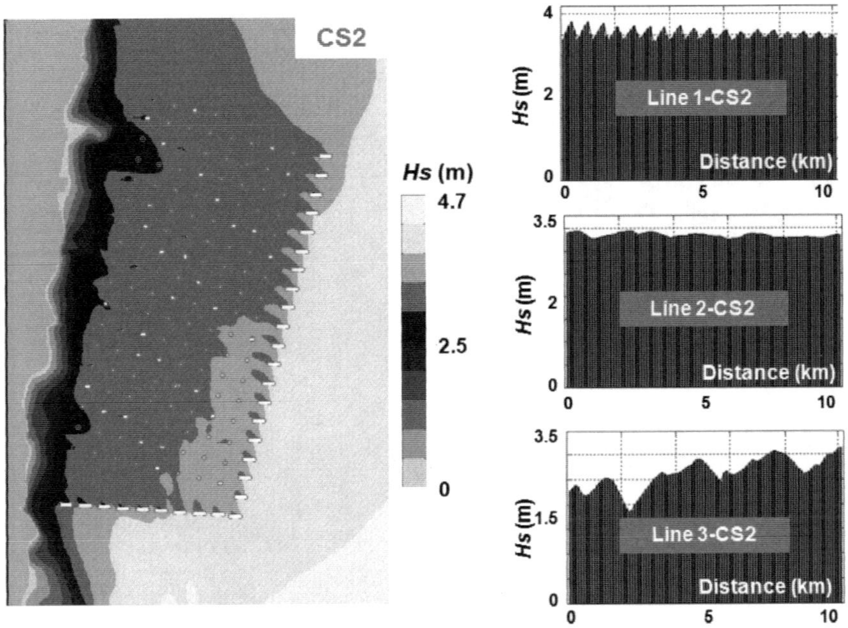

Figure 4. Evaluation in geographical space of the significant wave heights considering CS2 scenario.

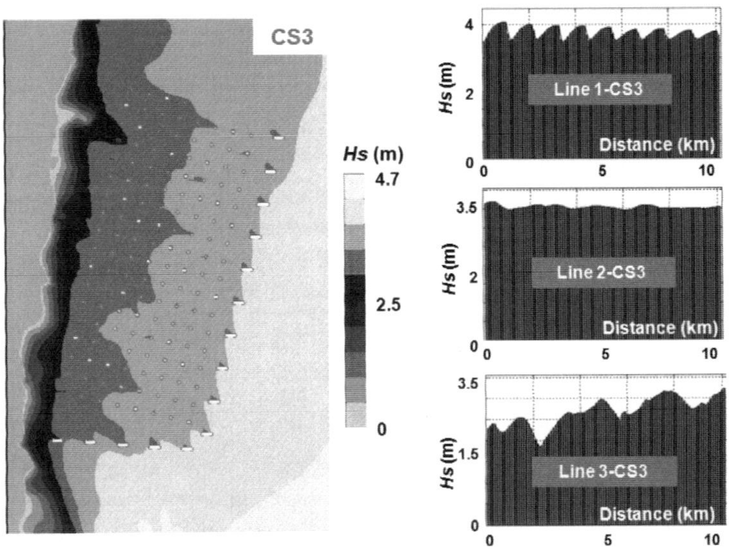

Figure 5. Evaluation in the geographical space of the significant wave heights considering the CS3 scenario.

A similar distribution is presented in Figure 6 for the CS4 case study, where two lines of WECs were taken into account. On a first evaluation it seems that the spatial impact of this configuration is similar to the one reported for CS2, however from the analysis of the line 1 profile we can noticed that the crest of the waves are trimmed which highlight the presence of the second line of WECs. For this case, the *Hs* values reported along the line 2 are located below the 3.5 m limit.

It is difficult to quantify the influence of the wind and wave farm by using only the spatial maps and references lines, so as a next step the information provided by the reference points P1-P9 will be evaluated in details. Figures 7, 8, 9 and 10 highlight an analysis, where the differences reported between the no farm situation and the case studies were indicated in percentages. A positive value reveal an attenuation of the wave parameter, while a negative sign reveal a reverse trend.

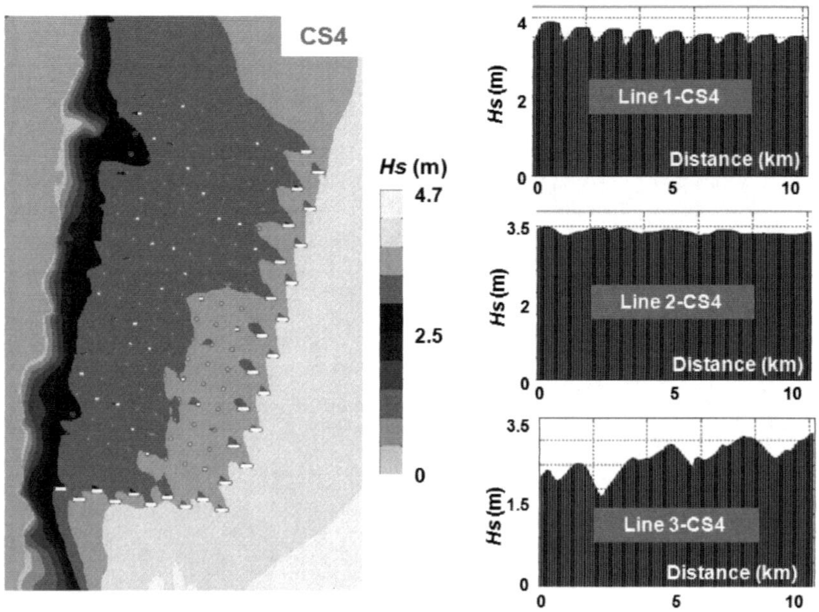

Figure 6. Evaluation in geographical space of the significant wave heights considering CS4 scenario.

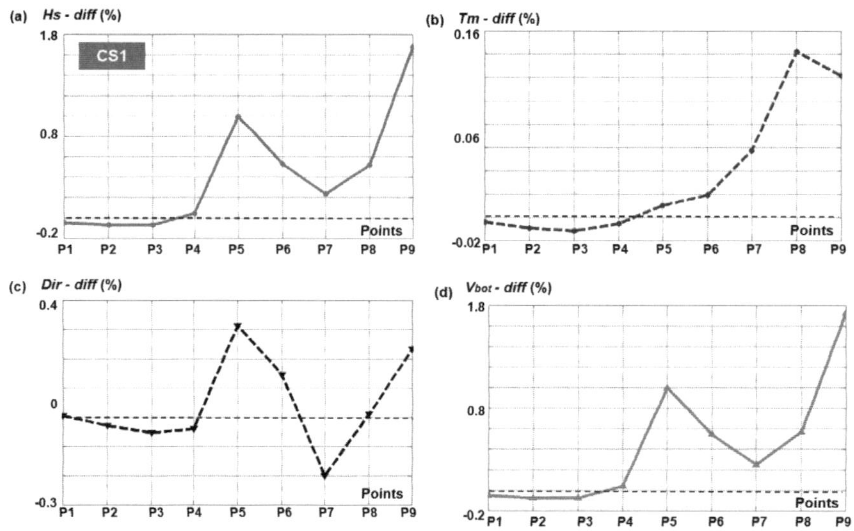

Figure 7. Variation of the wave parameters (in %) reported considering the no farm situation and the CS1 scenario, where: a) *Hs* variations; b) *Tm* variations; c) *Dir* variations; d) *Vbot* variations. The positive values indicate a decrease in magnitude in the presence of the farm, while the negative values indicate an opposite pattern.

https://doi.org/10.1051/e3sconf/20185101004

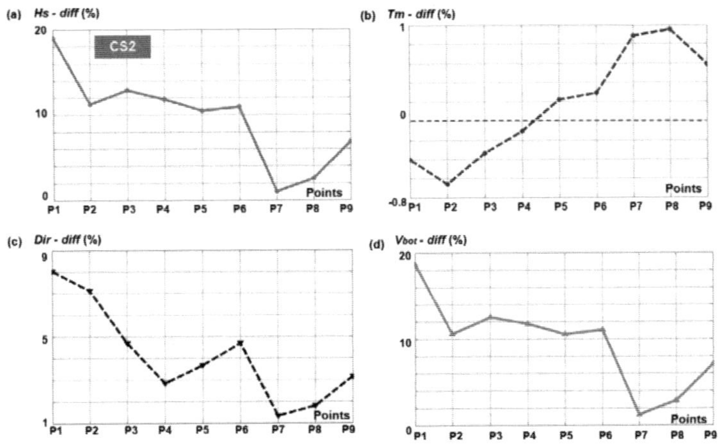

Figure 8. Variation of the wave parameters (in %) reported between the no farm situation and the CS2 scenario, where: a) *Hs* variations; b) *Tm* variations; c) *Dir* variations; d) *Vbot* variations.

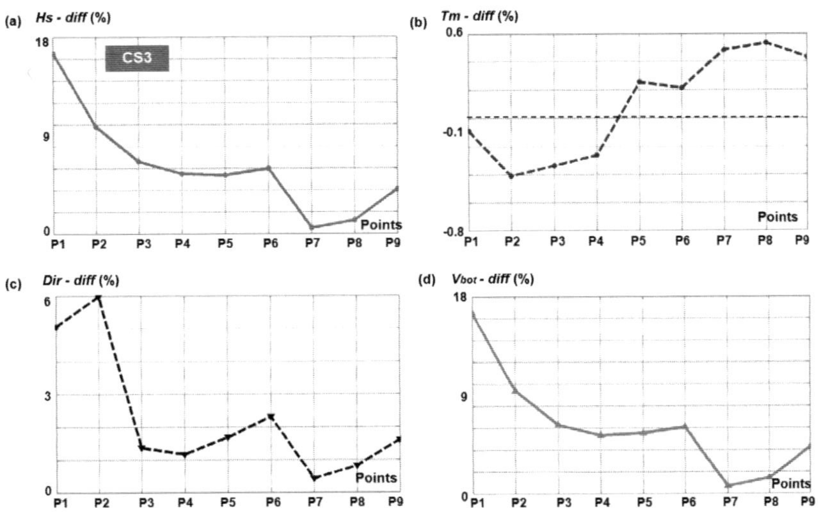

Figure 9. Variation of the wave parameters (in %) reported between the no farm situation and the CS3 scenario, where: a) *Hs* variations; b) *Tm* variations; c) *Dir* variations; d) *Vbot* variations.

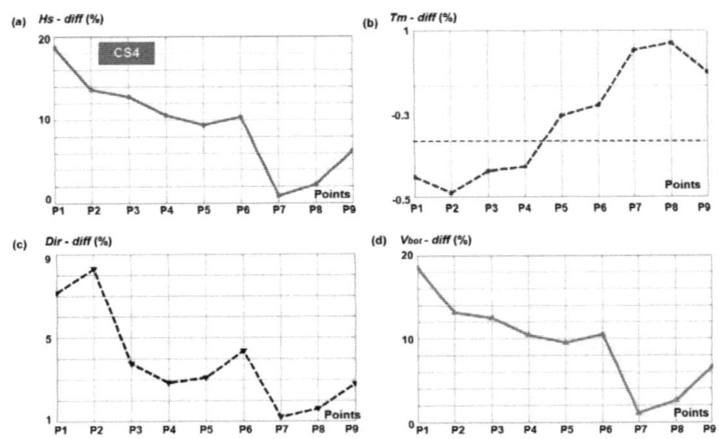

Figure 10. Variation of the wave parameters (in %) reported between the no farm situation and the CS4 scenario, where: a) *Hs* variations; b) *Tm* variations; c) *Dir* variations; d) *Vbot* variations.

4 Conclusions

The results presented in the present work highlight the coastal influence of several hybrid wind and wave farm configurations, which may operate in the vicinity of Saint George area, in the Romanian nearshore of the Black Sea. Several case studies were designed by using as a base an offshore wind farm, which has the same characteristics as the Greater Gabbard project from the UK, around which have been added several WECs deployed on a single or a two-line configuration.

An extreme scenario, associated with a winter storm, was considered for investigation and from the point of view of the operational limit of the workboats (set at 1.5 m) we can mention that no improvement was noticed in this direction, regardless of the case study considered for investigation. From the analysis of the spatial maps revealing the impact of marine farms in the target area, it was noticed that the shadow effect was induced only by the case studies CS2, CS3 and CS4, respectively.

Nevertheless, from the analysis of the values reported by the reference points P1-P9, it seems that only the presence of the offshore wind farm (the CS1 scenario) may influence the wave parameters, being reported a maximum 2% attenuation in the vicinity of the point P9 in the case of the parameters *Hs* and *Vbot*. As we go from the case CS1 to CS4, we can observe that the differences are more severe, being reported more consistent variations along the reference line 1, where a maximum value of 20% may be reported for the *Hs* parameter (CS2 scenario) compared to a 9% percentage reported for the *Dir* parameter (CS2 scenario). If the parameter *Vbot* decreases we expect to have an attenuation of the sediment movement since the bottom sediment will be less agitated.

Finally, it can be also mentioned that the work is ongoing, since at this moment there are no standard regulations regarding the layout of a wave farm, the main criteria being related to the optimal layout of the WECs in order to increase their energy efficiency. In general, these types of studies are focused only on the attenuation of the significant wave heights, but in fact by using multiple wave parameters it will be possible to provide a complete picture of the coastal impact induced by a marine energy farm.

Acknowledgements

This work was carried out in the framework of the project ACCWA (Assessment of the Climate Change effects on the WAve conditions in the Black Sea), supported by the Romanian Executive Agency for Higher Education, Research, Development and Innovation Funding-UEFISCDI, grant number PN-III-P4-ID-PCE-2016-0028. The ERA20C data used in this study have been obtained from the ECMWF data server.

References

1. G. Nastase, A. Serban, A.F. Nastase, G. Dragomir, A. I. Brezeanu, N.F. Iordan, Hydropower development in Romania. A review from its beginnings to the present. Renewable and Sustainable Energy Reviews **80**, 297-312 (2017).
2. G. Dragomir, A. Serban, G. Nastase, A.I. Brezeanu, Wind energy in Romania: A review from 2009 to 2016. Renewable and Sustainable Energy Reviews **64**, 129-143 (2016).
3. F. Onea, E. Rusu, Wind energy assessments along the Black Sea basin. Meteorological Applications **21(2)**, 316-329 (2014).
4. L. Rusu, F. Onea, The performance of some state-of-the-art wave energy converters in locations with the worldwide highest wave power. Renewable and Sustainable Energy Reviews **75**, 1348-1362 (2017)
5. F. Onea, L. Deleanu, L. Rusu, C. Georgescu, Evaluation of the wind energy potential along the Mediterranean Sea coasts. Energy Exploration & Exploitation **34(5)**, 766-792 (2016).
6. M. Sano, J.A. Jimenez, R. Medina, A. Stanica, A. Sanchez-Arcilla, I. Trumbic, The role of coastal setbacks in the context of coastal erosion and climate change. Ocean & Coastal Management **54(12)**, 943-950 (2011).
7. E. Rusu, F. Onea, Study on the influence of the distance to shore for a wave energy farm operating in the central part of the Portuguese nearshore. Energy Conversion and Management **114**, 209-223 (2016).
8. F. Onea, E. Rusu, The expected efficiency and coastal impact of a hybrid energy farm operating in the Portuguese nearshore. Energy **97**, 411-423 (2016).
9. F. Onea, L. Rusu, Coastal impact of a hybrid marine farm operating close to Sardinia Island. OCEANS'15 MTS/IEEE GENOVA 18-21 May 2015 Genova, Italy.
10. The SWAN team. SWAN User Manual, SWAN Cycle III version 41.20. Delft University of Technology (2017).
11. E. Rusu, D. Conley, E. Ferreira-Coelho, A hybrid framework for predicting waves and longshore currents. Journal of Marine Systems **69(1-2)**, 59-73 (2008).
12. L. Mee, SEAFISH – Project Ref: 10517. Complementary Benefits of Alternative Energy: Suitability of Offshore Wind Farms as Aquaculture Sites. Marine Institute University of Plymouth Plymouth *PL4 8AA* (2006).
13. M. Veigas, G. Iglesias, Potentials of a hybrid offshore farm for the island of Fuerteventura. Energy Conversion and Management **86**, 300-308 (2014).
14. A.S. Koraim. Hydraulic characteristics of pile-supported L-shaped bars used as a screen breakwater. Ocean Engineering **83**, 36–51(2014).
15. P. Poli, H. Hersbach, P. Berrisford, D. Dee, A. Simmons, P. Laloyaux, ERA-20C Deterministic, Report 20, 48 pp; European Centre for Medium-Range Weather Forecasts (2015).

16. S. Astariz, C. Perez-Collazo, J. Abanades, G. Iglesias, Hybrid wave and offshore wind farms: A comparative case study of co-located layouts. International Journal of Marine Energy **15**, 2–16 (2016).
17. I.Y. Georgiou, J. Schindler. Numerical simulation of waves and sediment transport along a transgressive barrier island. USGS, Scientific Investigations Report 2009–5252 (2009).

 https://doi.org/10.1051/e3sconf/20185101005

Prediction of Horizontal Axis Wind Turbine Rotor Performance: Bond Graph Approach

Naima Jouilel[1,*], Mohammed Radouani[1] and Benaissa El Fahime[1]

[1]ENSAM Meknes, Mechanical Department, Moulay Ismail University, Morocco

Abstract. Modeling wind energy conversion systems is a difficult task that requires the use of a unified language gathering all aspect of energies involved such as kinetic energy, mechanical energy, and electrical energy. Bond Graph methodology is an appropriate tool to analyze wind turbine dynamic behavior since the whole system is modelled in the same frame. Herein, a methodology for HAWT's rotor modeling is proposed based on Bond Graph, aerodynamic laws and Rayleigh Beam theory. It takes into consideration the profile, chord, and twist change along the blade. The model is validated using 20-Sim software and then compared to other models from literature. Simulation results show a better value of power coefficient in comparison with works using the same tools.

1 Introduction

Wind energy is extracted from the kinetic energy available in the wind via a conversion system which should be characterized by low environmental impact, low investment, and high yield power generation. This category of energy has recently experienced significant growth and is widely seen as a potential alternative to fight against the greenhouse effect. Many architectures were developed to extract energy from wind. Horizontal Axis Wind Turbine (HAWT) has demonstrated to be a transcendent tool in extracting the energy from wind and is characterized by a higher energy efficiency, less sensitiveness to off-design conditions, and more mechanical stability compared with Vertical Axis Wind Turbine (VAWT). Figure 1 shows HAWT's main subsystems in term of BG words.

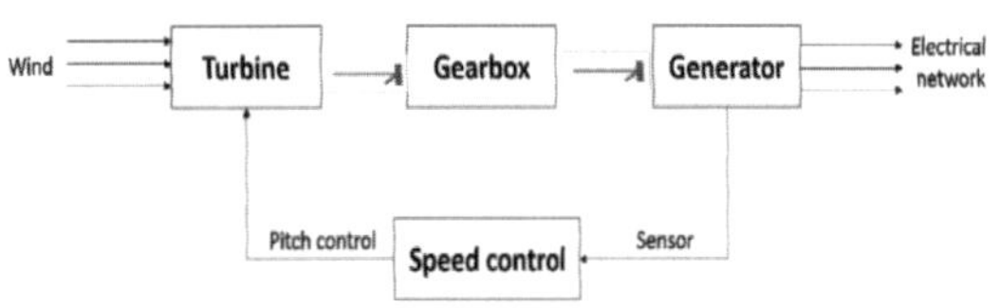

Figure 1. HAWT's main subsystems

The increase in wind generation remains limited. In fact, the maximum power that can be extracted is 60% due to Betz limit CP [1]. The fundamental equation (1) was first introduced by the German engineer Albert Betz in 1919 and published in his book "Wind Energy and its Extraction through Wind Mills" in 1926. The developed theory applies to both horizontal and vertical axis wind turbines.

$$C_P = 4a(1-a) \quad (1)$$

Several works have been conducted to develop wind turbines with performance that greatly approximate this limit [2–6], some of the studies refer to rotor dimensioning. In [7], bond graph model of wind turbine blade was performed using Rayleigh beam theory. In [8], bond graph model of wind turbine system: rotor, gearbox, and generator linked to the network was presented. Those models neglect some specificities of blade complex geometry. The aim of our work is to establish a detailed bond graph model of wind turbine blade considering profile, twist, and chord change along the blade in order to evaluate its efficiency. The proposed model has twofold objectives:

- Understand the rotor dynamic behavior for design purposes with regard to the numerous geometric specificities.
- Allow changes in introduced parameters to seek high power efficiency.

This paper is organized as follow: the second section introduces Bond Graph formalism used in this study and the main steps for HAWT's modeling proposed methodology. In the third section, the method presented in [9] is exploited for rotor geometric dimensioning and the blade element momentum (BEM) is used to establish aerodynamic model. The last section presents the proposed BG model for wind turbine rotor and a verification using 20-Sim software, results will be compared with data from literature for 750KW wind turbine.

2 BOND GRAPH FORMALISM FOR DESIGN MODELING

*Corresponding author: naima.jouilel@gmail.com

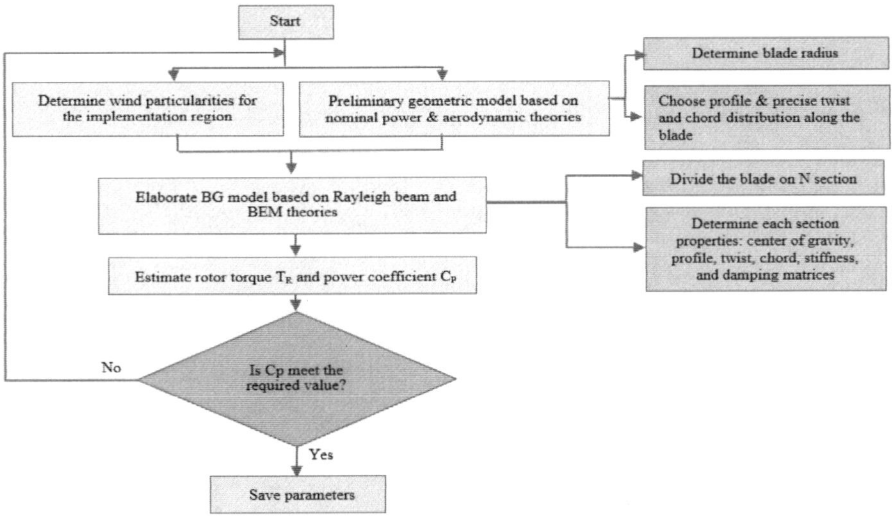

Figure 2. Flowchart of the proposed design optimization approach

2.1 Bond Graph methodology

Bond graph (BG) methodology for system modelling is based on the characterization of power exchange phenomena within a system [10,11].BG is a powerful tool in modelling systems; especially in cases where subsystems of different physical character (mechanical, electrical, thermal, hydraulic …) interact. The concept was introduced for the first time by Paynter in 1961 [12] and implicates the use of two energy variables: flow f (t) and efforts e (t), as shown in figure 2 (a). Each physical domain is associated to a type of flow and effort.

The adoption of a multi-discipline modeling tool is an asset in finding a good design solution. Thus, the use of BG formalism has increased significantly in last years; figure 2 (b) illustrates the evolution of using BG in recent researches according to Science Direct database for different research field. The significant use of this approach in multidisciplinary systems is due to its capacity to analyze the dynamic behavior of systems with high complexity.

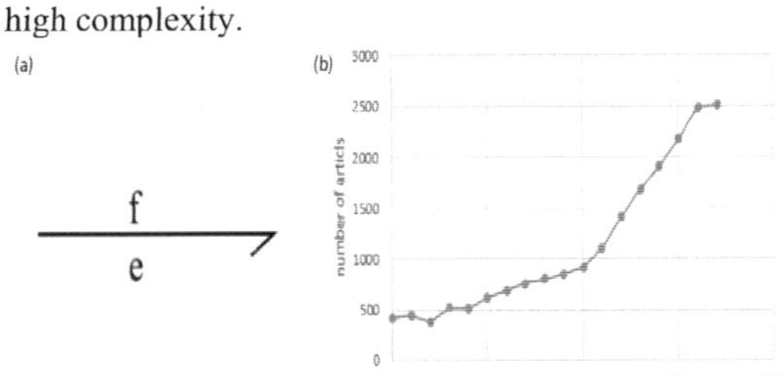

Figure 3. - (a) Link in a BG - (b) Evolution of Bond Graph use in recent researches

2.2 An aided design approach based on BG for HAWT's rotor modeling

The present work proposes a methodology for HAWT's rotor modeling in the design phase using BG methodology and aerodynamic theories. The flowchart of figure 2 resume basic steps on the proposed method. An application for 750 KW HAWT's rotor is displayed in the next section.

3 HAWT'S BLADE MODEL

3.1 GEOMETRIC & BOND GRAPH MODEL

In previous work [9], blade geometric model was performed based on a combination of the theories of axial flow and theory of blade member, then we looked for the NACA profile which provide the maximum blade efficiency compared to others. Herein, the parameters for a NACA 4415 are used to validate the proposed BG model. For aerodynamic efforts estimation, blade element momentum theory (BEM) was used [13-14], detailed in next section. Figure 4 shows geometric parameters results.

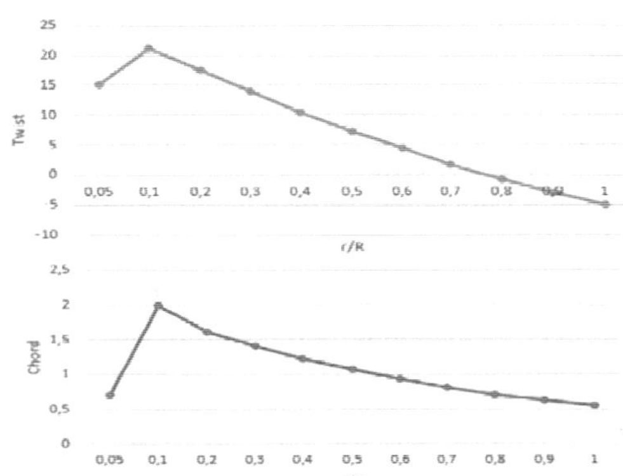

Figure 4. Chord and twist angle distribution

As shown in figure 5 (a), each rotor's blade is composed by two different profiles with different length. The first part has a circular profile (in direct connection with the hub) and the rest has a NACA profile with different chord distribution. Other blade's configurations propose more than one profile along the blade to ensure maximum efficiency [15].

For blade modeling with BG, "Rayleigh beam theory" was used [8]. This consists on considering the blade as a beam divided into N sections, each of which is subjected to efforts created by wind flow (figure 5 (c)). The following equation gives the relation that links Newtonians forces and displacements in each blade section:

$$\begin{bmatrix} F_i \\ M_i \\ F_{i+1} \\ M_{i+2} \end{bmatrix} = [K_i] \begin{bmatrix} y_i \\ \theta_i \\ y_{i+1} \\ \theta_{i+1} \end{bmatrix} \quad (2)$$

The stiffness matrix depends on the length of the section li and on the flexural stiffness EI according to relation (3). Figure (6) shows blade BG model in case we take three sections. The first one with a circular profile and the others with NACA 4415 profile. Sf_bound =0 represents a rigid connexion with the hub. Figure (7) illustrate the proposed model linked to hub and gearbox.

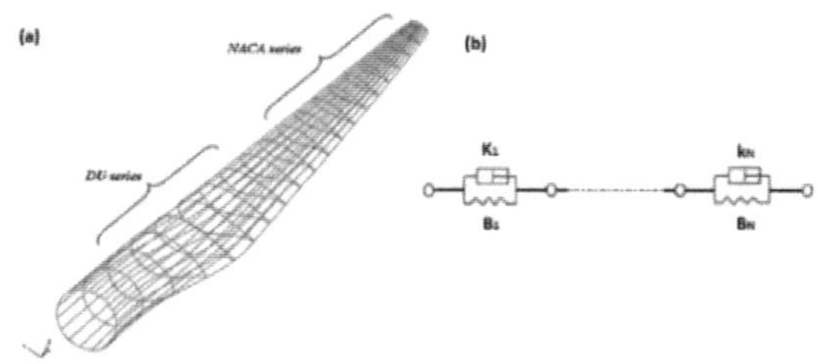

Figure 5. (a) Profile disposition along the blade (b) dynamic model by dividing the blade into N section (c) Newtonian efforts and displacements in a blade section

$$[K_i] = \frac{8EI}{(l_i + l_{i+1})^3} \begin{bmatrix} 12 & 6(l_i + l_{i+1})/2 & -12 & 6(l_i + l_{i+1})/2 \\ 6(l_i + l_{i+1})/2 & 4(l_i + l_{i+1})/2^2 & -6(l_i + l_{i+1})/2 & 2(l_i + l_{i+1})/2^2 \\ -12 & -6(l_i + l_{i+1})/2 & 12 & -6(l_i + l_{i+1})/2 \\ 6(l_i + l_{i+1})/2 & 2(l_i + l_{i+1})/2^2 & -6(l_i + l_{i+1})/2 & 4(l_i + l_{i+1})/2^2 \end{bmatrix} \quad (3)$$

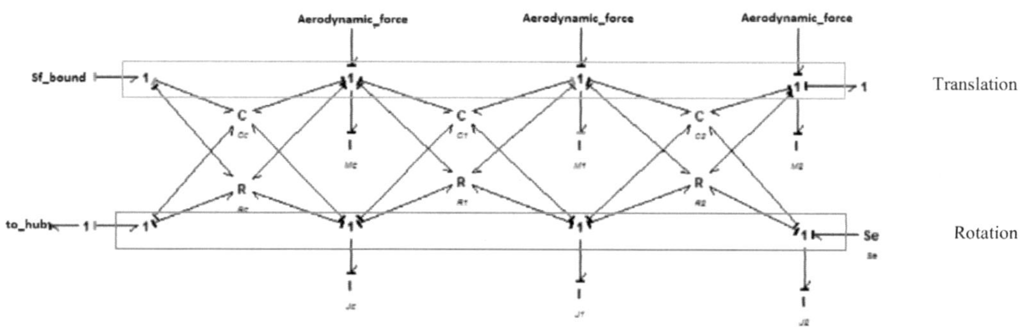

Figure 6. Blade model with three sections and two profiles

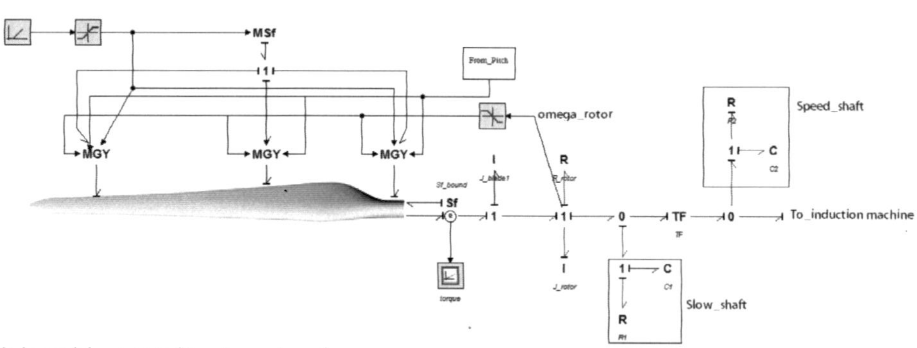

Figure 7. Blade model connected to rotor and gearbox

3.2 AERODYNAMIC MODEL

Aerodynamic modeling of wind turbines is an essential step in the design of this machine. Its goal is to calculate the aerodynamic forces exerted on the blades to determine blade's optimal parameters and estimate the maximum power extracted by the rotor. This modeling is a complex task because of the rotation of the helix which creates a three-dimensional flow of air around the blades. The theory of BEM applied in this work is used in the calculation of aerodynamic forces acting on the wind turbine blades. This theory makes use of several parameters, in particular those related to the blade geometry and the used profile (figure 8).
The lift and drag forces can be resolved into axial and tangential components. The former is called axial thrust force and can be expressed as:

$$F_{ti} = \left(\frac{1}{2}\rho V_w \frac{(1-a_i)^2}{sin^2\phi_i}(C_{li}\ sin\phi_i - C_{di}cos\phi_i)C_i l_i\right)V_w = \mathrm{M}\ \mathrm{V_W} \quad (4)$$

In equation (4), M represents the gyrator's modulus, which is used for the conversion of the wind energy into aerodynamic energy using modulated gyrator (MGY).

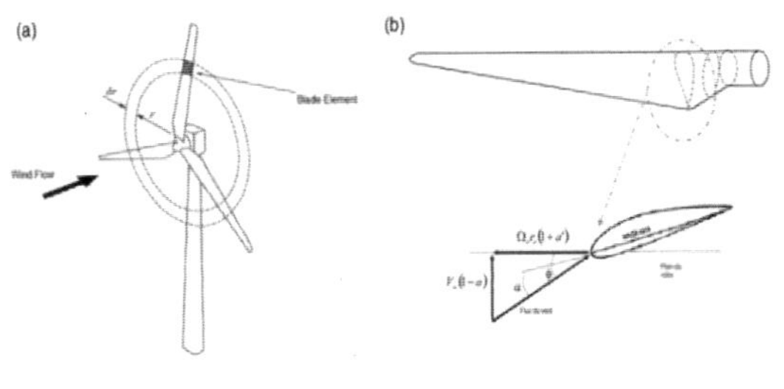

Figure 8. (a) Wind turbine rotor; (b) Projected velocities in the plane of the rotor

Both lift and drag forces contribute to the axial thrust force. Further, the lift force develops useful torque whereas the drag opposes it. Therefore, a high ratio CL/CD is desired to achieve high conversion efficiency. Axial interference factor and tangential interference factor are calculated using mathematical code, based on BEM theory, for NACA4415 profile.

With $$\boldsymbol{\phi_i = \tan^{-1}\left(\frac{V_w(1-a_i)}{\Omega_r r_i(1+a_i')}\right)} \quad (5)$$

$$\boldsymbol{a_i = \left(1 + \frac{4sin^2\phi_i}{\sigma_i'(C_{li}cos\phi_i + C_{di}sin\phi_i)}\right)^{-1}} \quad (6)$$

$$\boldsymbol{a_i' = \left(-1 + \frac{4sin^2\phi_i}{\sigma_i'(C_{li}cos\phi_i - C_{di}sin\phi_i)}\right)^{-1}} \quad (7)$$

4 RESULTS & DISCUSSION

In order to analyse dynamic responses of the proposed model, two different simulations were carried out. The first one is launched by introducing a constant wind speed. In the second one, variable wind was introduced using the assumption that wind speed never exceeds its higher limit which is 22 m/s, it allows the system to work without pitch control. Simulation results show an increase in power coefficient: $C_P \approx 0.4147$ compared to [8] where Cp was about 0.33, with 25,6% as difference percentage for the same HAWT rotor. The variable wind submodel is generated with random values ϵ [5 m/s, 22 m/s], this allows to test the model responses following a speed change in wind flow.

The system undergoes some oscillations in the beginning of the simulation due to the dynamic behaviors in the transient phase. The total useful torque developed by the turbine depends on a dimensionless coefficient CQ (figure 11) and it is expressed as:

For $\boldsymbol{\lambda}$=5, VW=9 m/s and Cp=0.41 as given below, we have Tr=139337 N.m. This theoretical value matches the obtained one by model simulation, where Tr=3*42487=127461 N.m with a gap of ~8.5 %. It is mandatory to mention that the focus on this study is the blade geometric specificities. Therefore, if the pitch angle model and other wind turbine subsystems are added, the total useful torque will take its average limit.

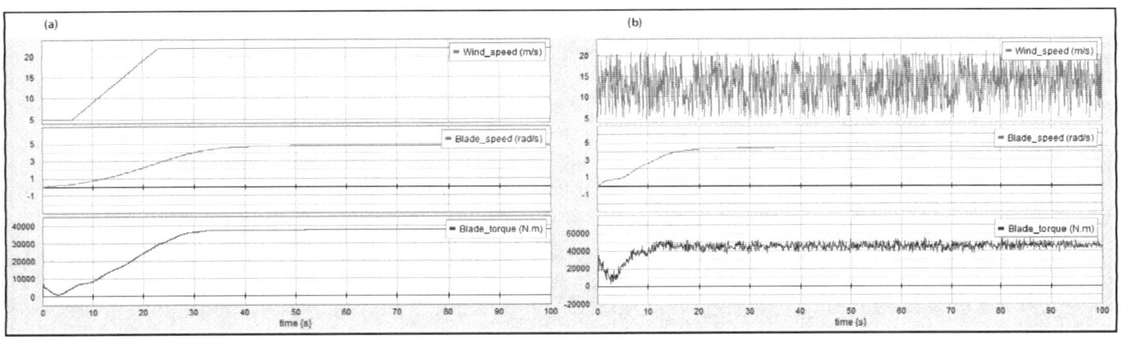

Figure 9. Simulation results (a) for constant wind and (b) variable wind speed

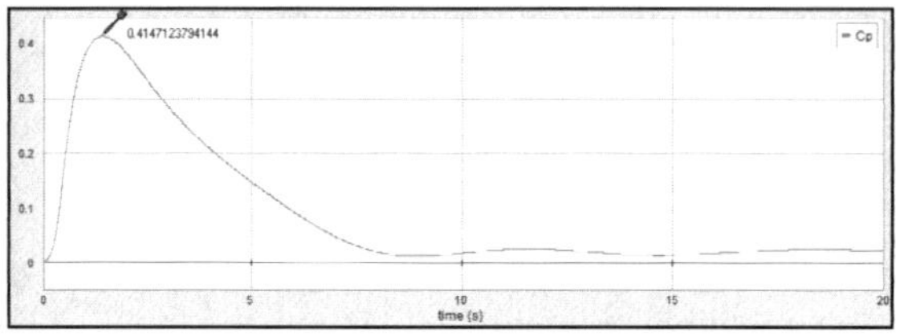

Figure 10. Power coefficient curve

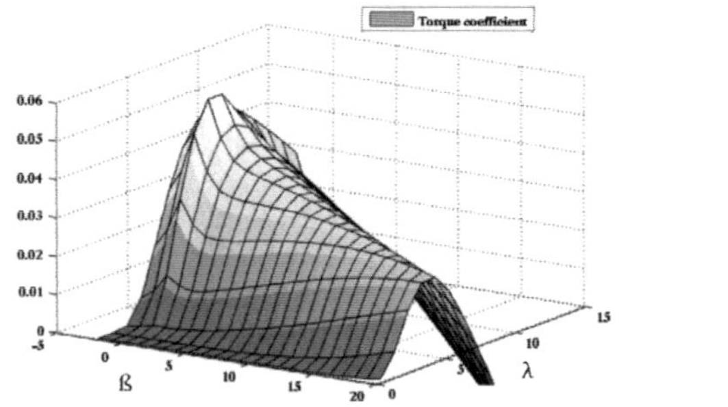

Figure 11. Torque coefficient as a function of pitch angle β and tip speed ratio λ

5 CONCLUSION

Being a particular multidisciplinary system, a mechatronic approach for HAWT modeling is necessary. Thus, BG methodology is proposed as an innovative and versatile tool for the study of this system. The model realization using BG approach allows the estimation of several parameters facilitating the prediction of the system dynamic behavior and efficiency. Furthermore, the uses of this formalism are diversified to meet the different modeling needs such as optimization, generation of states equations, and determination of robust control laws by inverse BG.

In this work, a model of wind turbine blades based on BG has been developed. The proposed model is based on several parameters combining the aerodynamic laws and the blades complex geometry. A rotor performance prediction for a 750 KW wind turbine was made for two simulation cases: constant and variable wind speed. The steps of this study constitute a contribution to parameters design and optimization. Thus, it makes possible to estimate the variation effect of the geometrical features on the rotor response. Among these parameters, we mention the profile variation that was omitted in previous works.

In order to better predict the rotor responses, an extension of this study by adding other subsystems is envisaged in future papers. The important subsystem whose model must be detailed is the source of the system: the wind, which is characterized by its intermittency and its deterministic (wind shear, tower shadow) and stochastic parameters (rotational turbulence). On the other hand, an addition of pitch control model will increase the rotor efficiency by tracking the optimum angle of attack to reach and maintain the wind turbine nominal power.

TABLE I. NOMENCLATURE

a	Axial interference factor	C_Q	Torque coefficient	R	Blade radius
a'	Tangential interference factor	Ω	Rotor speed	V_w	Air speed
Ci	Section chord	F_{ti}	Thrust force	l_i	Section length
C_L	Lift coefficient	λ	Specific speed (Tip speed ratio)	ϕ	flow angle
C_D	Optimum value of the drag coefficient	λ_r	Local specific speed	α	Incidence angle
C_P	power coefficient	r	Relative radius	β	Optimal angle of twist

TABLE II. SIMULATION PARAMETERS

Blade structure & aerodynamics		Section 1	Section 2	Section 3	Rotor		
li	Section lenght (m)	1.6	11.7	11.7	E	Young modulus (N/m^2)	17^{e}9
C_i	Chord (m)	1	1.9	1	μ	Structural damping coefficient	0.01
$ß_i$	Twist angle (deg)	0	7.234	5.678	D_{hub}	Main bearing (N/m)	1000
Z_i	Center of gravity position (m)	0.8	7.5	18.3	J_{hub}	Hub inertia (Kg.m^2)	5000
I_i	Moment of inertia (Kg.m^2)	3e10-5	6e10-5	6e10-5	R	Rotor Radius (m)	22
Mi	Mass of section (Kg)	70	1000	450			
ρ	Air density(Kg/m^3)		1.225				

https://doi.org/10.1051/e3sconf/20185101005

References

1. J. M. Jonkman, "Modeling of the UAE Wind Turbine for Refinement of FAST AD Modeling," 2003.
2. D. Afungchui, B. Kammoun, and A. Chauvin, "Development of a Wind Turbine Blade Profile Analysis Code Based on the Singularities Method," in *Wind power*, 2003, pp. 81–88.
3. R. Belamadi, A. Djemili, A. Ilinca, and R. Mdouki, "Aerodynamic performance analysis of slotted airfoils for application to wind turbine blades," *J. Wind Eng. Ind. Aerodyn.*, vol. 151, pp. 79–99, 2016.
4. F. G. Jin Chen , Quan Wang, Shiqiang Zhang, Peter Eecen, "A new direct design method of wind turbine airfoils and wind tunnel experiment," *Appl. Math. Model. ELSEVIER*, vol. 40, pp. 2002–2014, 2015.
5. C. Thumthae, "Optimum Blade Profiles for a Variable-Speed Wind Turbine in Low Wind Area," *Energy Procedia*, vol. 75, pp. 651–657, 2015.
6. J. A. S. Witteveen, S. Sarkar, and H. Bijl, "Modeling physical uncertainties in dynamic stall induced fluid-structure interaction of turbine blades using arbitrary polynomial chaos," *Comput. Struct.*, vol. 85, no. 11–14, pp. 866–878, 2007.
7. G. AgarwalS, ChalalL, Dauphin-TanguyG, "Bond graph model of wind turbine blade.," *Bond graph Model. Pract.*, 2012.
8. R. Sanchez and A. Medina, "Wind turbine model simulation : A bond graph approach," *Simul. Model. Pract. Theory*, vol. 41, pp. 28–45, 2014.
9. J. Naima, R. Mohammed, and Elf. Benaissa, "Three-dimensional modeling of a horizontal axis wind turbine blade and profile effect analysis," in *IEEE International renewbal and sustainbale energy conference,* 2016, pp. 1–5.
10. Wolfgang Borutzky, *Bond Graph modelling of engineering systems*. springer, 2011.
11. Wolfgang Borutzky, *Bond Graph methodology: Development and Analysis of Multi- disciplinary Dynamic System Models*. Springer, 2010.
12. HM Paynter, "Analysis and design of engineering systems." The M.I.T. Press, Cambridge, 1961.
13. M. O. L. Hansen, *Aerodynamics of Wind Turbines* .
14. Z. Sun, J. Chen, W. Z. Shen, and W. J. Zhu, "Improved blade element momentum theory for wind turbine aerodynamic computations," *Renew. Energy*, vol. 96, pp. 824–831, 2016.
15. Y. J. Lee, Y. T. Jhan, and C. H. Chung, "Fluid-structure interaction of FRP wind turbine blades under aerodynamic effect," *Compos. Part B Eng.*, vol. 43, no. 5, pp. 2180–2191, 2012.

Prediction of the wave power in the Black Sea based on wind speed using artificial neural networks

Sorin Ciortan[1,*], Eugen Rusu[1]

[1]EMechanical Engineering Department, Engineering Faculty, "Dunarea de Jos" University of Galati, Galati, Domneasca Street, 111, Galati, Romania

Abstract. The paper proposes a prediction methodology for the significant wave height (and implicitly the wave power), based on the artificial neural networks. The proposed approach takes as input data the wind speed values recorded for different time periods. The prediction of significant wave height is useful both for assessment of wave energy as also for marine equipment design and navigation. The data used cover the time interval 1999 to 2007 and it was measured on Gloria drilling unit, which operates in the Romanian nearshore of the Black Sea at about 500 meters depth.

1 Introduction

It is well known that the wave's height value is important information needed for the activities that intend to recover the energy from waves. The power that a wave can deliver is directly linked [1] to the wave's height. Taking into account that this value is influenced by a lot of factors like wind speed, water temperature, geographic location, date and time, etc., is difficult to predict its evolution [2]. There are out there mathematical models dedicated to this job [3] capable to deliver fairly precise results but there is also room for other ways to do such predictions.

In order to predict wave's height, the main working way is to find all possible influences and render these into mathematical formulas. In the real world, there are so many influences on wave behaviour that even most complex the models can miss some.

Another aspect of wave prediction is linked to the data used for. In order to build a model some previous data must be computed. The model itself will simulate the links between the data and will work only for datasets which have same links between. As anyone can observe, in the late years the wheatear condition are continuously and quickly changing [4]. So, models built on datasets ten, even five, years old can give less precise results for newer ones.

The paper try to prove that a prediction model based on Artificial Neural Networks (ANNs) is suitable for wave prediction if the dataset used are relatively new and more, that for every dataset used for training there is an optimal network architecture.

The ANN based models are neither unique nor new [5] and offer the advantage of "black box" simulation. The influences of all factors taken into account are not transposed into mathematical equations but are implemented "as is" into neural network itself. This way, difficult or even impossible to mathematically simulate interactions can be included.

As target example, the wave height prediction near Gloria (Romanian marine platform) located in the Black Sea, figure 1, was used. In this area, the conditions are appropriate for wave energy acquiring [6] but, also for extreme phenomenon [7].

We try to prove that an ANN based model can be used for significant wave height prediction, but the basic data sets must be related to the domain of weather modifications.

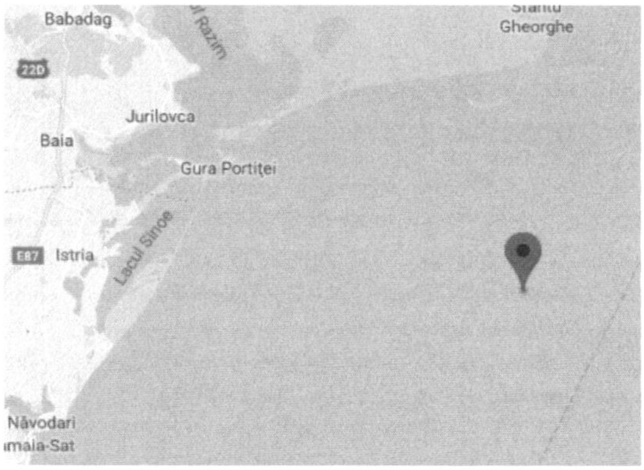

Figure 1. Romanian marine platform location [*https://maps.google.com*]

2 Artificial Neural Network based modeling

Artificial Neural Networks are parallel calculus systems that overcome some disadvantages of sequential calculus machines (base on von Newman architecture) like:

* Corresponding author: sorin.ciortan@ugal.ro

 https://doi.org/10.1051/e3sconf/20185101006

calculus brake if missing data, compulsory computing algorithm, stop at errors etc.

Basically, an ANN simulate a biological brain, both as architecture and working system. Being composed by many elementary units (so called "artificial neuron") distributed in several areas - layers, [8], figure 2. The ANN can accumulate information, through a "learning or training procedure". During this stage, some known input-output datasets are presented to the ANN and internal links between the neurons are adjusted until the desired results are obtained.

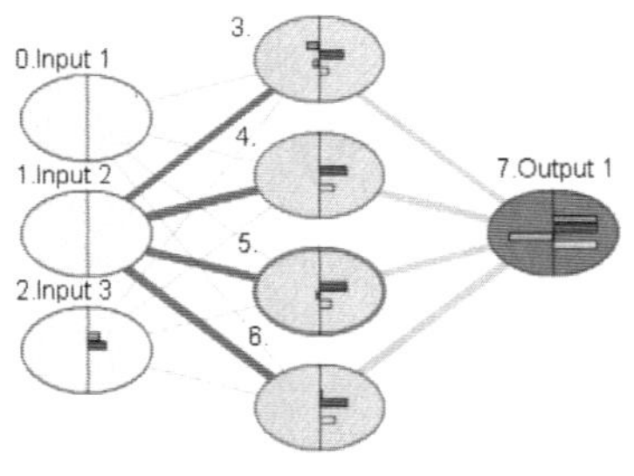

Figure 2. General Feed-forward neural network architecture

After the training procedure is finished, the ANN can be used with several goals: prediction - to find unknown output data for known input data, optimization - to find optimal input values for desired output or analysis - to find which input data has most influence on outputs.

There are several types of ANNs, with different architectures and specific data flows, for this work a feed-forward ANN trained with back-propagation algorithm was chosen. Since for every simulated phenomenon the ANN has an optimal architecture (number of neurons and layers), several variants must be tested in order to establish the best one.

ANN based models have also a drawback: the precision is less comparing with a sequential calculus. So, from the beginning, when using a neural model, an error is assumed. Anyway, there are situations when this fact is less important than the other facilities offered.

There are several dedicated software, allowing to create, train and interrogate ANNs in order to obtain predictions. Among these, EasyNN offers all necessary facilities that present work needs.

2.1 Dataset pre-processing

In order to be used for ANN modeling, the datasets, must be fitted in appropriate files (tab separated text), one for each year. Due to large number of values (approx. 1450 / year), just one decimal place was kept for wind and wave values. This option facilitates the learning process of the ANN model, taking into account the big differences in wave height over a year [9,10]. In figure 3 is presented, as an example, the wave values for 2007.

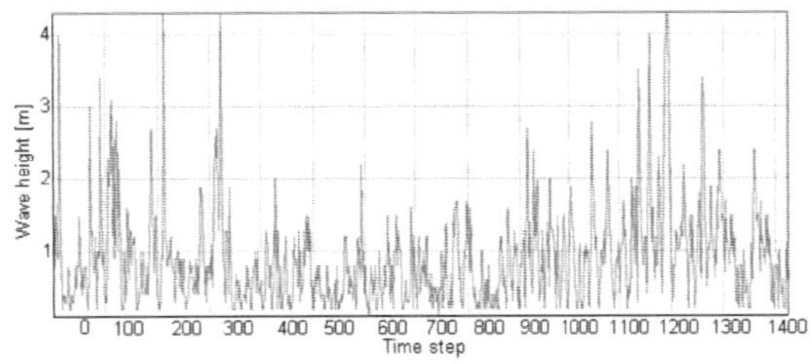

Figure 3. Wave height evolution over year 2007.

2.2 ANN building and training

As the establishing of the network architecture represents the most important stage in the ANN model building, an optimization procedure was used: several networks were built and trained for a low number of cycles and the best was chosen. EasyNN has a built-in genetic algorithm based module which performs this task. Taking into account that the wind have the main influence on wave height [11], as training datasets, the values for wind speed and significant wave height were used. In order to cover all the connections between the wind and wave, the datasets contain values at every day/month, for 1999-2006, with 6 hours time step. The date, time and wind speed stand as inputs and wave height as output. As results, eight ANNs were obtained, one for each year data-set, named further as M-1999....M-2007.

For training procedure, a low learning rate was used, in order to avoid oscillation of the process. As training error the value of 0.1 was targeted. The details are presented in table 1.

Table 1. The characteristics of the ANN models.

ANN code	Structure	Training cycles	Training error	Learning rate	Momentum
M-1999	4-4-4-1	531899	0.09	0.3	0.8
M-2000	4-4-4-1	469031	0.09	0.3	0.8
M-2001	4-4-4-1	124793	0.09	0.3	0.8
M-2002	4-7-5-1	122713	0.09	0.3	0.8
M-2003	4-8-4-1	263235	0.09	0.3	0.8
M-2004	4-8-4-1	40914	0.08	0.3	0.8
M-2005	4-8-4-1	345667	0.09	0.3	0.8
M-2006	4-9-3-1	48664	0.08	0.3	0.8

As can be observed in table1, there are differences between the architectures along the years 1999 - 2006. These differences lead to the conclusion that there are some significant changes between the data sets. Usually, an ANN becomes more complex - high number of neurons - when the data are more scattered. This may occur due to climate modification from 1999 to 2006.

In figure 4 it is presented, as an example, the ANN architecture for two different years.

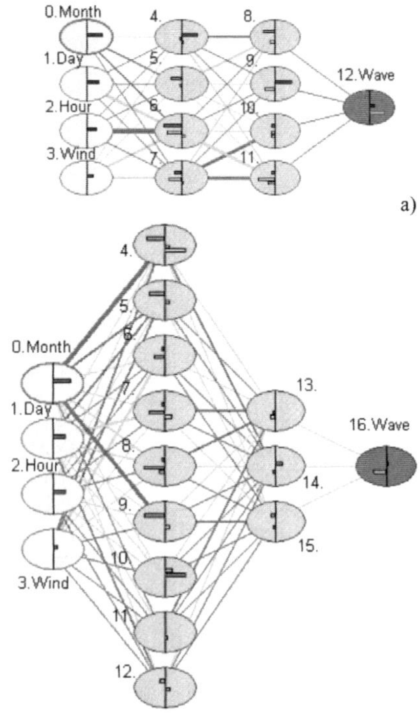

a) b)

Figure 4. ANN architectures

a) 1999 dataset; b) 2006 dataset

2.3 ANN model predictions

With the trained models, the prediction phase can be approached. With this goal, every year dataset was used as input for previous years ANN models and the prediction results were compared with the actual wave height. As significant parameter, the RMSE was used, computed from each time step value for every year. The corresponding results are presented in table 2.

Table 2. ANN models prediction comparison

	Prediction precision (RMSE)							
	2007	2006	2005	2004	2003	2002	2001	2000
M-2006	0.190	---	---	---	---	---	---	---
M-2005	0.239	0.172	---	---	---	---	---	---
M-2004	0.196	0.189	0.251	---	---	---	---	---
M-2003	0.171	0.190	0.214	0.174	---	---	---	---
M-2002	0.186	0.191	0.249	0.198	0.239	---	---	---
M-2001	0.175	0.210	0.258	0.213	0.240	0.201	---	---
M-2000	0.163	0.211	0.248	0.181	0.265	0.199	0.198	---
M-1999	0.185	0.201	0.253	0.220	0.257	0.225	0.221	0.211

3 Discussion

Looking to the results presented in table 2, one can observe that RMSE values are different for different ANN models and, in most cases, these are increasing with the oldness of dataset used for training. This is in well concordance with the hypothesis that during the time the climate conditions are constantly changing. This change modifies the wave height evolution and older models, trained with old datasets gets low in prediction precision. This change is obvious also in graphical representations of wave evolutions, figure 5.

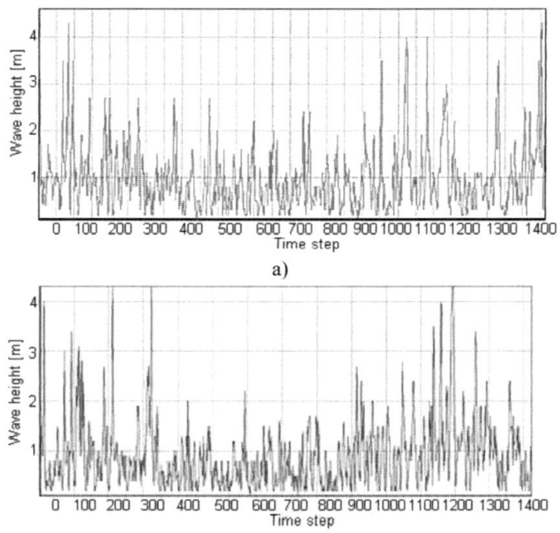

a) b)

Figure 5. Wave height evolution comparison

a) year 2000 b) year 2007

Another aspect that must be taken into account is the percent of low significant wave height values, comparable to the high ones, present in the training dataset. According to [12], an ANN, during the training stage, will try to emulate the values with greater presence in dataset: low, medium of high. This leads to higher prediction errors when the actual conditions are favourable to other values distribution than data set training. This phenomenon is more obvious in case of low times step values, when height peaks are more likely to be recorded.

For the ANN models here were used datasets covering one whole year, with six hours as value for time step, as a consequence the high wave peaks, due to storms, are present. From this point of view, the obtained RMSE values are consistent with those obtained, in similar conditions, on other datasets [12,13].

4 Conclusions

Following all presented above, some conclusions can be drawn:

- The ANN based models can be used successfully for wave height prediction, allowing the wave energy extraction optimization.

- The predictions precision is directly influenced by the datasets used for ANN training, i.e. time step, number of values and high-low values proportion.

- Another factor of influence on prediction precision is the oldness of training datasets: as these are recent as the prediction is more precise.
- After choosing the appropriate training dataset, the most important step in ANN model building is represented by establishing the architecture. In this purpose a module based on genetic algorithms is highly recommended.

The work is still ongoing and different approaches and datasets are considered.

ACKNOWLEDGEMENTS

This work was carried out in the framework of the research project REMARC (Renewable Energy extraction in MArine environment and its Coastal impact), supported by the Romanian Executive Agency for Higher Education, Research, Development, and Innovation Funding – UEFISCDI, grant number PN-III-P4IDPCE-2016-0017.

References

1. E. Rusu, Energies **7**, 4002-4018 (2014)
2. C. Gasparotti E. Rusu, J. of Env. Prot. and Eco. **13**, 1751-1759 (2012)
3. F. Onea, S. Ciortan E. Rusu, En.& Env. **28**, 580-597 (2017)
4. L. Rusu,D. Butunoiu, E. Rusu, J. of Env. Prot. and Eco. **15,** 445-454 (2014)
5. O. Makarynskyy, D. Makarynska,E. Rusu Mar. Transp. and Exp. of Oc. and Coas. **1 2,** 1085-1091 (2005)
6. E. Rusu, J. of Mar. Sc. and Tech. **14,** 359-372 (2009)
7. V. Covalenco, S. Ciortan, E. Rusu Mech. Test. & Diag. **2** 16-23 (2017)
8. S. Haykin *Neural networks, a comprehensive foundation* (MacMillan College Publishing Co., New York, 1994)
9. E. Rusu, Energies **7**, 4002-4018 (2014)
10. F. Onea, A. Raileanu, E. Rusu, *Advances in Meteorology* (2015)
11. F. Onea, E. Rusu, *Met. App.* **21** 316-329 (2014)
12. S.N. Londhe, V. Panchang, J. of Atm. and Oc. tech. **23** 1593-1603(2006)
13. B. Paplinska-Swerpel L. Paszke Arch. of Hydro-Eng. and Env. Mech. **53** 183-201 (2006)

An overview of the wind power potential in the Romanian coastal environment - moving from onshore to offshore

Dragos M. Niculescu[1,2,*], and Eugen V. C. Rusu[1]

[1]Department of Mechanical Engineering, University of Galati „Dunarea de Jos", 111 Domneasca Street, Galati, Romania
[2] Department of Oceanography, National Institute for Marine Research and Development "Grigore Antipa", 300 Mamaia Blvd., 900581, Constanța, Romania

Abstract. In the past few years in Romania the total installed capacity reached 3025MW. Because the Europe Union is pursuing to obtain an increase in renewable energy by 2020 to cover 20% of the energy consumption with renewable energy maybe in the near future in Romania the wind energy parks will start to be build off shore where the wind energy is more powerful. From this perspective, the present work is trying to provide a better picture to the wind energy resources at the Romanian shore. The measured data was provided by NOAA [1] from different points (weather stations) four along the Romanian shore and one off-shore. The measured data was compared with data from the Global Wind Atlas model [2]. According to the analysis, there are discrepancies between the measured results and those provided by the model, but the two sets of data show that in the northern part of the off-shore the power density and the wind speed is higher than in the south.

1 Introduction

The study area is situated in the black sea at the Romanian coast, see fig. 1. The sea it's an enclosed basin and it could be considered to be one of the most distant extension of the ocean, being linked to the Atlantic Ocean via three straits (Gibraltar, Dardanelles and Bosporus) and two seas (the Mediterranean and Marmara).

Figure 1. The Black Sea basin in the summer season – Sentinel 2 (the red rectangle encompasses the study area).

Even if it is not a very large sea, many strong storms occurs in this area (see for example Rusu, 2015 and 2016) [3,4], in fact many sailors are not fond of the Black Sea because when a storm hits they have no were to go to avoid the heavy swell (one might describe it like being in a large pool). There's a real risk for marine and coastal hazards when big storms occurs (Gasparotti and Rusu, 2012) [5]. The heaviest afflicted part of the Black Sea is its westerner region were the underwater topography is the shallowest (Rusu et al., 2014) [6] and the coasts are subjected to a dynamic erosion process while others are subject to accretion processes. Furthermore, the north-western part of the Black Sea is where the Danube River flows into the sea creating the Danube Delta and where the process of wave-current interaction are very strong with a direct influence on the neighboring coastal dynamics (Ivan et al., 2012)[7]. Many coastal works need to be carefully planned, as in some cases they might appear to be beneficial at first sight, but after a long period of time they prove to be of no help with an unwanted outcome (Omer I. et al. 2015 and Vlasceanu E. et al. 2015) [8,9].

Based on recent studies more and more attention is paid to the off-shore wind regime near the Romanian coast aiming to explain or comprehend the wind conditions as a ground for future studies regarding the hotspots for installing wind farms (Onea F. et al. 2014, 2015, 2016) [10,11,12].

2 Methodology

It should be assumed that the wind speeds between the on-shore and off-shore data will differ on some degree as the roughness factor of the sea is rather small in comparison with the roughness factor on the land. Even so because the two data sets come from different outputs (e.g. measured and weather model) the data can't be compared as it is. The data sets were analyzed from different perspectives to manage a correlation and contribute in a small degree to the understanding of how the wind speed varies when scaling it in the model.
To calculate the wind power density for the measured data (from meteorological stations) the following equation (1) was used.

*Corresponding author: n.dragos8@gmail.com

$$P_{wind} = \frac{\rho_{air} \cdot U_x^3}{2} \quad (1)$$

ρ_{air} – represents the air density (≈1.22 kg/m^3)

U_x – the wind speed reported by meteorological station

2.1 Description of the study

The study aims to make a correlation between the measured data and the model data. The model data are taken from the "Global Wind Atlas" while the measured data were taken from NOAA. The weather model uses the best available global datasets and the result validation is done with the help of remote sensing data from Synthetic Aperture Radar (SAR) (Alfredo Peña et al., 2015) [13]. The power density for the measured wind speed was calculated and it was put alongside with the interpolated model results.

2.2 Onshore wind – measured

The data used in preliminary estimations of the energetic potential were realized, corresponding to some series of data obtained from several weather stations across the shore line and one inland. The estimations are annual and there results (Table 1) show the average of annual energy potential and the standard deviation. To be noted that the M. Kogalniceanu weather station is fairly in land approximately 16 km, in open field, as a result of this the power distribution is the smallest. The other three stations are on the coast, thus having grater power densities (Niculescu and Rusu 2016) [14].

Table 1. Preliminary estimations for the energetic potential

Weather Station	Yearly average (W/m^2)	Standard Deviation	Wind Speed Average
Sulina	207.9	57.2	6.21
M. Kogalniceanu	67.1	14.5	4.44
Constanta	239.6	62.5	3.58
Mangalia	83.3	27.7	3.41

For Sulina, Constanta and Mangalia the height at which the data is collected is as follows 9m, 14m and respectively 9m. The height of M. Koglaniceanu weather station is unknown (Cristescu T. M., 2015) [15].

Nonetheless, using the scaling formulations denoted by the Monin-Obukhov scales (2) the height of the station can be approximated to 9m. A graphical depiction can be seen in Figure 2.

Friction velocity $u_* = \overline{-u'w'}$

Temperature scale $T_* = -\frac{\overline{-w'\theta}}{u_*}$ (2)

Water vapour scale $q_* = -\frac{\overline{-w'q'}}{u_*}$

Monin-Obukhov stability length $L = \frac{T \cdot u_*^2}{kg\left(T_* + 0.61 \cdot q_*\right)}$

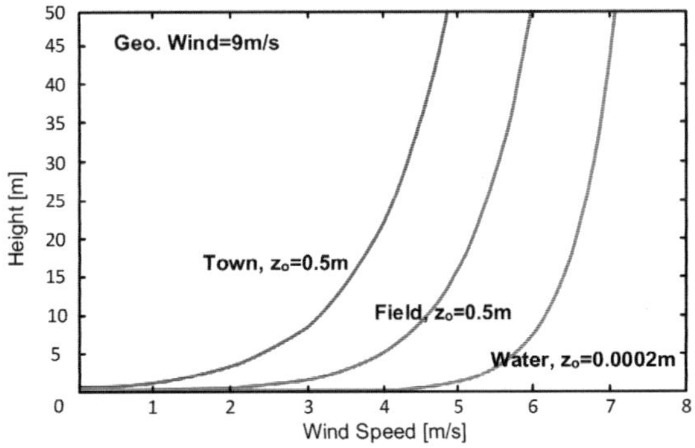

Figure 2. The change in surface wind for different roughness values, with the geostrophic wind, G = 9m/s.

2.3 Offshore Wind – measured

The offshore data was collected on an oil rig between 1995 and 2007 and has almost 85000 values resulting in a wind speed average of 7.97 m/s (Table 2) at a height of approximately 28m (Cristescu T. M., 2015) [15].

Table 2. Preliminary estimations for the energetic potential

Weather Station	Annual average (W/m^2)	Standard Deviation	Wind Speed Average
Gloria(ANM)	412.1	19	7.97

2.4 Offshore Wind – data resulted from the model

It is well known that the behavior of the atmosphere is in a continuous motion, therefore it's not easy to characterize it nonetheless it can be done through the temporal periodicities and horizontal spatial scales (micro-scale, mezzo-scale and macro-scale). All this periods and scales are tied to the energy change, together contributing to the temporal variability in wind energy density in a specific location (Barthelmie R. et al., 2009) [16].

The data used in this study is a result of "The Global Wind Atlas" a free web-based application (Global Wind Atlas). The considered data points can be found in Fig. 3 (a) 50m, (b) 100m and (c) 200m altitude (standard elevation given by the model). The total number of points considered is 17.

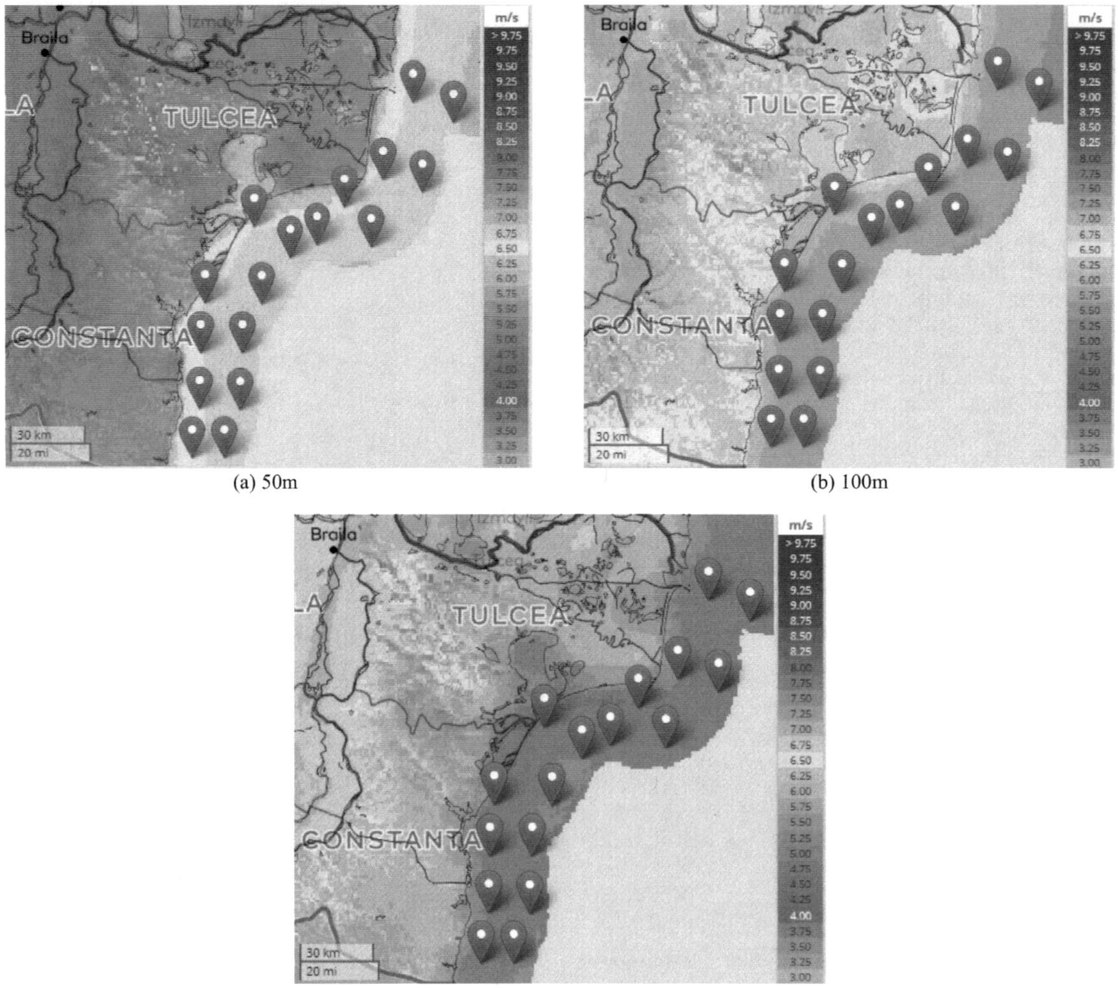

(a) 50m (b) 100m

(c) 200m

Figure 3. Data collection points at thee elevation

3 Results

The measured on-shore and off-shore data sets were used alongside with the_model results to obtain the interpolated result as a map using the Kriging Interpolation method.

The red diamonds represents in figure 4 the on-shore points of the weather stations, the black triangle represents the oil rig (Gloria Oil Rig) and the green dots point from where the model results were selected.

Analyzing the interpolation results between the measured data and the data results of the weather model, several particularities can be noted. First of all, the weather station data are measured at different heights than the height of the model output results. This in turn would make the measured data not relevant for the interpolation. But when noting the average wind speed that was registered at the Gloria Oil Rig at the height of 28m, a question can be asked: Why the wind speed is greater at this altitude than the models result at 50m? One of the answers would be that the model boundary has a limitation over the seas and oceans and because this value is at the limit it may not be correct.

The wind speed value from Sulina weather station is fairly close to the value that would correspond when estimating the values with the help of the scaling formulations, still the value is lower than the models result value of 6.61 near it. This in turn can point out that when considering an average on long wind data sets (temperature or water vapour) the state of the atmosphere is thermally neutral.

The values for Constanta and Mangalia are significantly lower; a possible explanation is that the respective meteorological stations are located in areas where there are shading effects due to post-site construction. By analysing the three maps (Figure 5) witch were a result of the interpolated model data it can be noticed how the wind velocities increase in the north part. In the third image (200m height) the interpolation result shows three hot spots where the wind speed is between 7.69 and 7.75.

https://doi.org/10.1051/e3sconf/20185101007

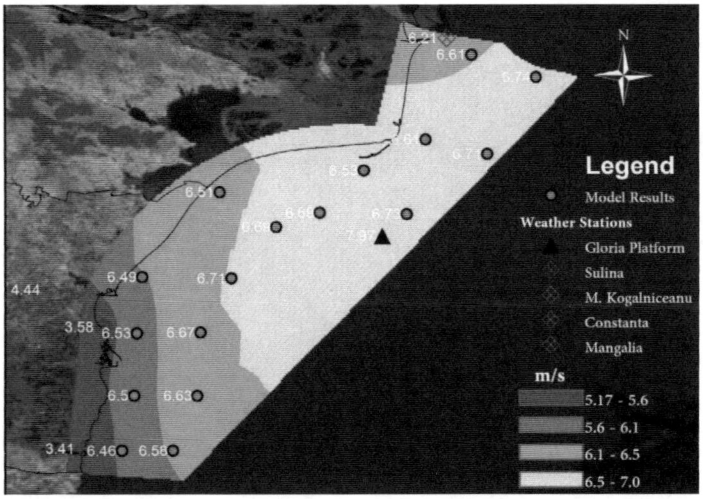

Figure 4. Wind Speed interpolation.

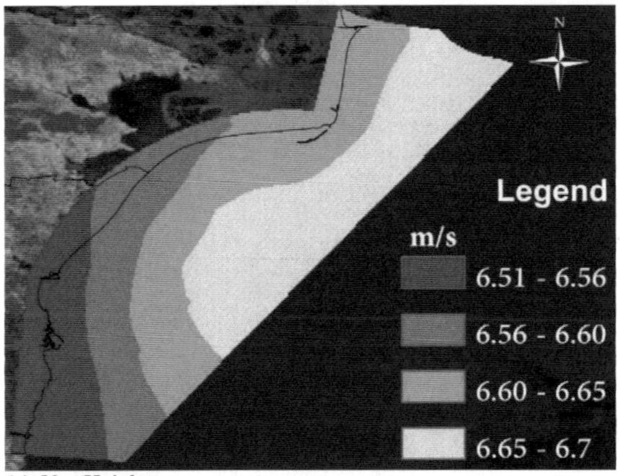

(a) 50m Height

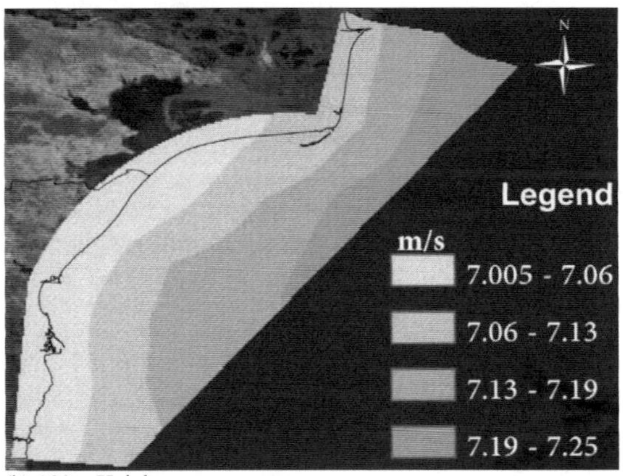

(b) 100m Height

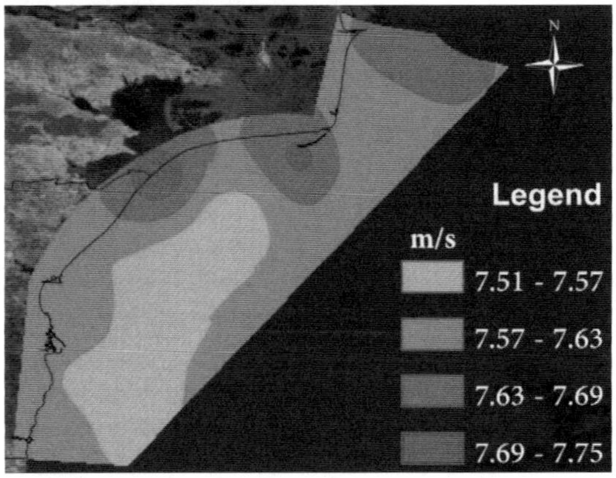

(c) 200m Height

Figure 5. Wind Speed at different heights.

The interpolation of the power density data is shown in Figure 6. Based on the interpolated result, the mainland influence can be observed along the shore. In the southern part the topography is high with cliffs while in the north the orography is low do to the Danube Delta. As in the case above (Figure 4), there are differences between the measured data and model data. The value of Gloria oil rig weather station (412.1 W / m2) is higher by 14.8% compared to the nearest value of the model (350.9 W/m2) and it should be noted that the elevation of the weather station is lower that the model's.

A short comparison between the three images in figure 7 show how the interpolation result follows more or less the shore line. In the first image (a) 50m the pattern of the interpolation mimics the shoreline. In the second image (b) 100m the interpolation pattern is straighter and the maximum power density moves a bit to the north. At the height of 200m the maximum power density is still in the northern part of the coast but the interpolation pattern is wavy. This phenomenon might be liked with the large scale topography (the mountain regions that surrounds the aria creating a wind corridor).

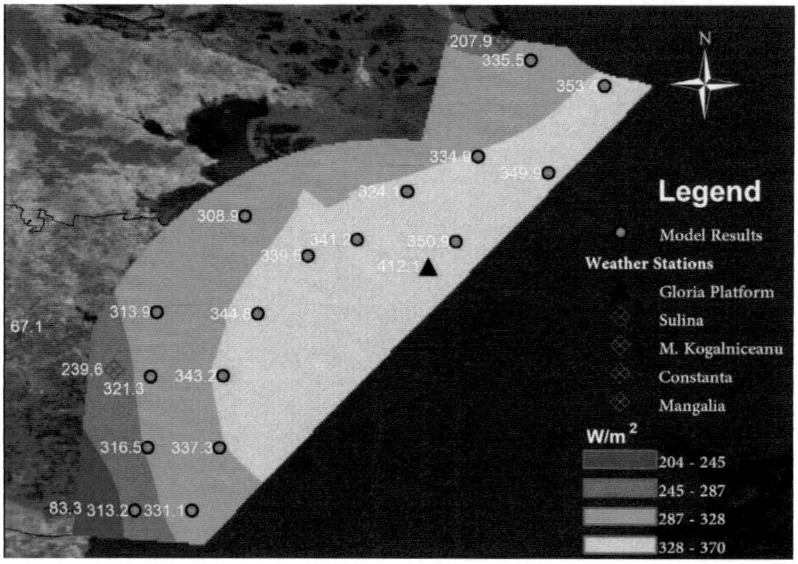

Figure 6. Power Density interpolation

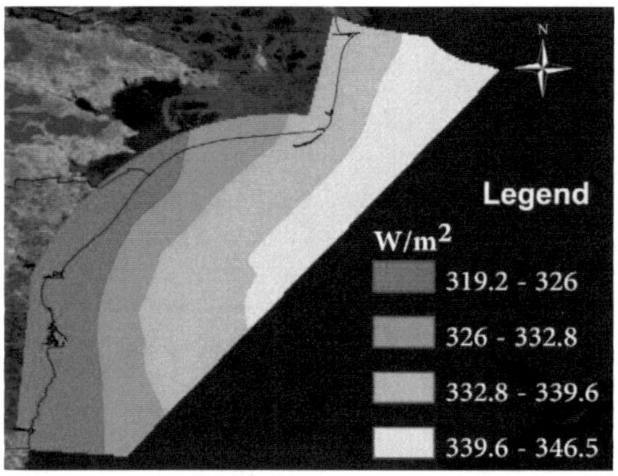

(a) 50m Height

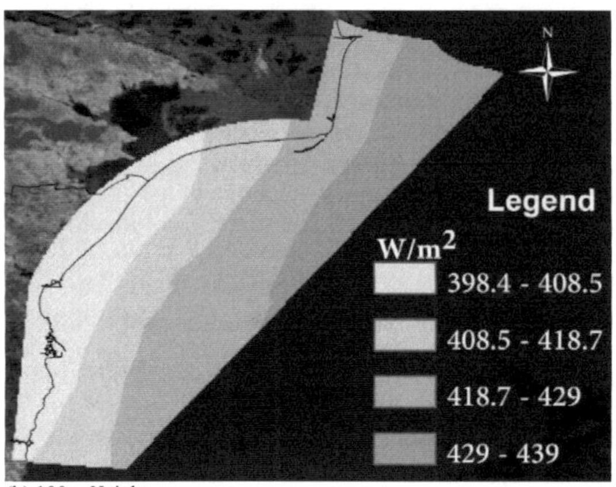

(b) 100m Height

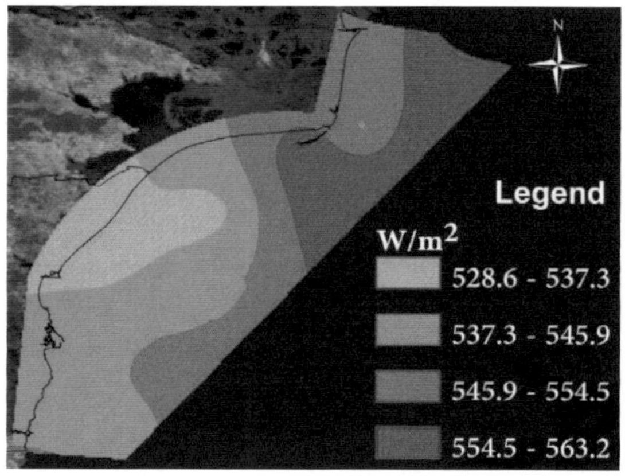

(c) 200m Height

Figure 7. Power Density at different heights.

4 Conclusions

Based on the current study the power densities near the Romanian coastal environment are not close to the offshore power potential of some other coasts, as those from the northern seas where the wind speeds and power densities are far higher.

Nonetheless, when considering the Romanian coastal wind energy potential there might be a chance considering that the wind speed at the Gloria oil rig the mean value is 7.97m/s and the power density is 412W/m2. Furthermore, the models output data and measured data might suggest that in the northern part of the Romanian near shore, the power density is much higher than in the southern part.

The work is still ongoing and some further studies and data collection campaigns should be conducted in order to validate the numerical models against the measured data.

Acknowledgments

This work was carried out in the framework of the research project REMARC (Renewable Energy extraction in MARine environment and its Coastal impact), grant number PN-III-P4-IDPCE-2016-0017, financed by the Romanian Executive Agency for Higher Education, Research, Development and Innovation Funding – UEFISCDI.

References

1. NOAA National Centers for Environmental Information, https://gis.ncdc.noaa.gov/map/viewer/#app=cd
2. Technical University of Denmark, (DTU Wind Energy), https://globalwindatlas.info/
3. Rusu, E., MATEC, 2016; 62. EDP Sciences.
4. Rusu, L., Energies, 2015; **8**(9), 10370-10388.
5. Gasparotti, C. and Rusu, E. JEPE, 2012; **13**(3-A), 1751-1759.
6. Rusu, L., Butunoiu, D. and Rusu, E., JEPE, 2014; **15** (2), 445-454.
7. Ivan, A., Gasparotti, C. and Rusu, E., JEPE, 2012; **13**, 1673-1682.
8. Omer, I., Mateescu, R., Rusu, L., Niculescu, D. and Vlăsceanu, E., JEPE, 2015; **16**(2), 417-424.
9. Vlasceanu, E., Niculescu, D., Omer I. Rusu, E. and Ivan, A., SGEM Conference Proceedings, 2015; Book3 **2**, 675-682.
10. F Onea, E Rusu. Meteorological Applications, 2014; **21** (2), 316-329
11. Onea, F.; Raileanu, A. and Rusu, E., Advances in Meteorology, vol. 2015; ID 808617, 14 pages.
12. Onea, F. and Rusu, E., Energy Exploration & Exploitation, 2016; **34**(2), 217-234.
13. Peña, A., Hasager, C. B., Lange, J., Anger, J., Badger, M., Bingöl, F., Würth, I. DTU Wind Energy E; No. 0029(EN)); 2013.
14. Niculescu, D. and Rusu, E., MTD, 2016; **2**, 25-34.
15. Cristescu,T.M., Cercetari Marine, 2015; **45**,160-172;
16. Barthelmie, R & Pryor, Sara & Frandsen, S.T. Offshore Wind Power, 2009; 43-70.

Economic performance evaluation of natural gas vehicles and their fuel infrastructures

Dejene A Hagos[1,*] and Erik O Ahlgren[1]

[1]Department of Space, Earth and Environment, Chalmers University of Technology, SE-412 96 Göteborg, Sweden

Abstract. The transition from high carbon-intensity to low carbon-intensity transport fuels entails the development of energy efficient and cost-effective decarbonisation pathways. In this paper, 14 potential natural and renewable gas supply pathways and natural gas vehicles (NGVs) have been selected and evaluated with regards to well-to-tank (WTT) fuel production costs and break-even vehicle added investment costs. NGVs are evaluated for both road- and maritime transport applications with three types of gas engines; dedicated, dual fuel, and high pressure direct injection (HPDI) engines. The results indicate that owing to the alternate gas distribution mechanisms and filling stations configuration there exist a substantial fuel production cost differences between the selected gas pathways. Despite its long-distance shipping and distribution, imported LNG showed significant production cost advantage over compressed natural gas (CNG) and liquefied renewable natural gas (LRNG) pathways. Evaluating the current economic performances, all NGVs are found to be competitive corresponding to gasoline cars, but not compared to diesel cars due to the lower price gap between CNG and diesel. In the heavy-duty vehicle and passenger vessel segments, however, owing to the high price gap between LNG and diesel/marine gas oil (MGO), all NGVs and LNG passenger vessels showed high competitiveness compared to their conventional counterparts.

1 Introduction

In the last few decades, the transition to a low carbon economy and a green energy system have been a goal for many countries. The EU-27 countries reduced their combined total greenhouse gas (GHG) emissions during 1990-2014 by 23%. However, in the same period, the transport sector's emission increased by 20.1% [1]. Road transport took the greatest share, accounting for 73% of the total emission in 2014. This attributes to the higher emission mitigation costs of alternative transport systems. Efforts are being focused on the development of alternative fuel vehicles, improving vehicle power train system, and development of advanced vehicle fuels. As transition fuels, in recent decades, the use of natural gas (NG) and upgraded biogas or renewable natural gas (RNG) as vehicle fuels has become increasingly important. A high energy-to-carbon ratio, stable and low price, and abundant availability, and its clear burning characteristics makes NG an attractive alternative transport fuel. It could also potentially pave way for increased use of RNG. As a clear burning fuel, NG could also potentially reduce air pollutant emissions in urban areas. Nevertheless, its low energy density (which limits its driving range) and low cetane number (which restricts its use in compression engines without pilot diesel injection) are technical limitations. To increase its energy density and provide a longer driving range for NGVs, NG should either be compressed to about 200 bars and stored in high-pressure tanks (CNG), or cooled to -162°C at atmospheric pressure and stored in highly insulated cryogenic tanks (LNG). Globally, as of 2016, NGVs make up 1.32% of the total vehicle population [2]. NGVs are widely used in Asia Pacific and Latin American countries. In Europe, few countries have shown high market penetration of NGVs, such as Italy and Germany [3].

In maritime transport applications, especially for short-range passenger ferries, LNG is becoming important as alternative to expensive low sulphur marine fuels to meet the tight emissions standard in emission control areas (ECA), such as Baltic Sea area [4]. As of 2015, globally, more than 70 LNG ships were in operation; mainly regional ferries located in Norway. Nevertheless, lack of bunkering standards and regulatory framework, reduced cargo space, methane leakage (during bunkering), methane slip (unburned methane), availability of bunkering ports, and high upfront vessel added investment cost are main challenges facing the market [5]. In this regard, assessment of the fuel

*Corresponding author: dejene@chalmers.se

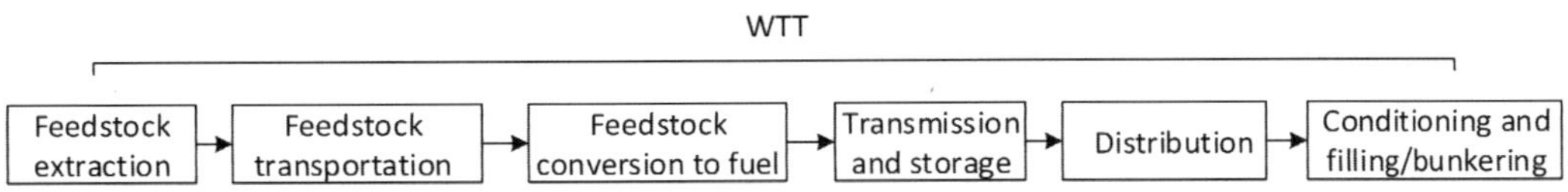

Figure 1. Major stages of the WTT cost evaluation.

production and conditioning costs of gas infrastructures and the added gas vehicle investment cost are critical to identify the true economic values of gas in transport, and for cost effective decarbonisation of the transport system at large. In most techno-economic modelling studies, due to the computational and modelling framework limitations, the detailed cost components of alternative fuels and vehicles often ignored or not properly captured. Thus, the aim of this study is to analyse the WTT fuel production and distribution costs of selected gas supply pathways and break-even added investment costs of gas vehicles, both for road transport application and maritime transport applications, within the Danish NG and RNG perspectives.

2 Methodology

In this section, the general approach, selected fuel supply pathways, the mode and types of gas vehicles included, and data sources and assumptions are discussed in brief.

2.1 General approach

The focus is on Denmark as a case. The assumed domestic feedstocks are the Danish NG mix (as of 2016), organic municipal waste, and cattle and pig manure, for biogas production. Figure 1 shows the major processes and stages of the evaluation, from feedstocks extraction to filling NGVs at filling stations and bunkering LNG passenger vessels at ports. The cost evaluations have been done using a model developed in Excel and comprise levelised fuel production costs at each stage, gas distribution costs to filling stations, and break-even added investment costs of NGVs. A general discount rate of 6% is assumed for all cases. The results of the levelised fuel production, distribution, and filling/bunkering costs estimations are reported in terms of €/GJ (2016 price level).

The break-even added cost is the added investment cost of a gas vehicle over a similar conventional vehicle (in terms of engine brake power) at which the net present added cost equals the net present benefit of the gas vehicle due to the fuel cost savings over the analysis period. The net present value (NPV) is given as :

$$NPV = \sum_{t=0}^{n} \frac{C_{NGV,t}}{(1+d)^t} - \sum_{t=0}^{n} \frac{C_{CV,t}}{(1+d)^t}$$

At break-even, or no-loss no-profit point, NPV=0

$$IC_{NGV} + \sum_{t=1}^{n} \frac{C_{NGV,t}}{(1+d)^t} = IC_{CV} + \sum_{t=1}^{n} \frac{C_{CV,t}}{(1+d)^t}$$

$$IC_{NGV} - IC_{CV} = \sum_{t=1}^{n} \frac{C_{NGV,t}}{(1+d)^t} - \sum_{t=1}^{n} \frac{C_{CV,t}}{(1+d)^t}$$

Where, IC_{NGV} and IC_{CV} refers to the investment costs of a specific NGV and the corresponding conventional vehicle, respectively while C_{NGV} and C_{CV} refers to the respective annual costs thereof, d refers to the assumed discount rate (6%), and t refers to the analysis period; assumed to be 13 years for cars, 7 years for HD vehicles, and 25 years for vessels.

The costs on the right-hand side include replacement costs, operation and maintenance costs, and salvage values, for each vehicle group. Then, the break-even added investment cost was calculated through an iterative process by varying the vehicle added cost (IC_{NGV}-IC_{CV}) until the net present added cost equals to the net present benefit of the gas vehicle (due to fuel cost savings over the analysis period), or simply, when both sides cost components are equal. This is because some of the cost components like replacement cost and salvage value are dependent on the vehicle added cost.

2.2 Selected fuel supply pathways

The pathway selection has been done, mainly, based on the authors' preceding state-of-the art review. The pathways are primarily classified based on filling station configurations and gas distribution mechanisms. In the context of Danish gas infrastructures, the most probable pathways are generated. Figure 2 shows the three major blocks of pathways based on the filling station configurations; mother, daughter, and LNG/LRNG stations; mother stations are grid connected, and gas is transported to by pipeline while daughter stations are off grid, and CNG is transported to by CNG swap body/trailer. Within each block, the pathways are classified based on availability (private vs. public) and filling time (fast fill vs. time fill); the main difference between fast-fill (5 minutes) and time-fill (6-8 hours) stations is the availability of large-size storage and compressors and their cost.

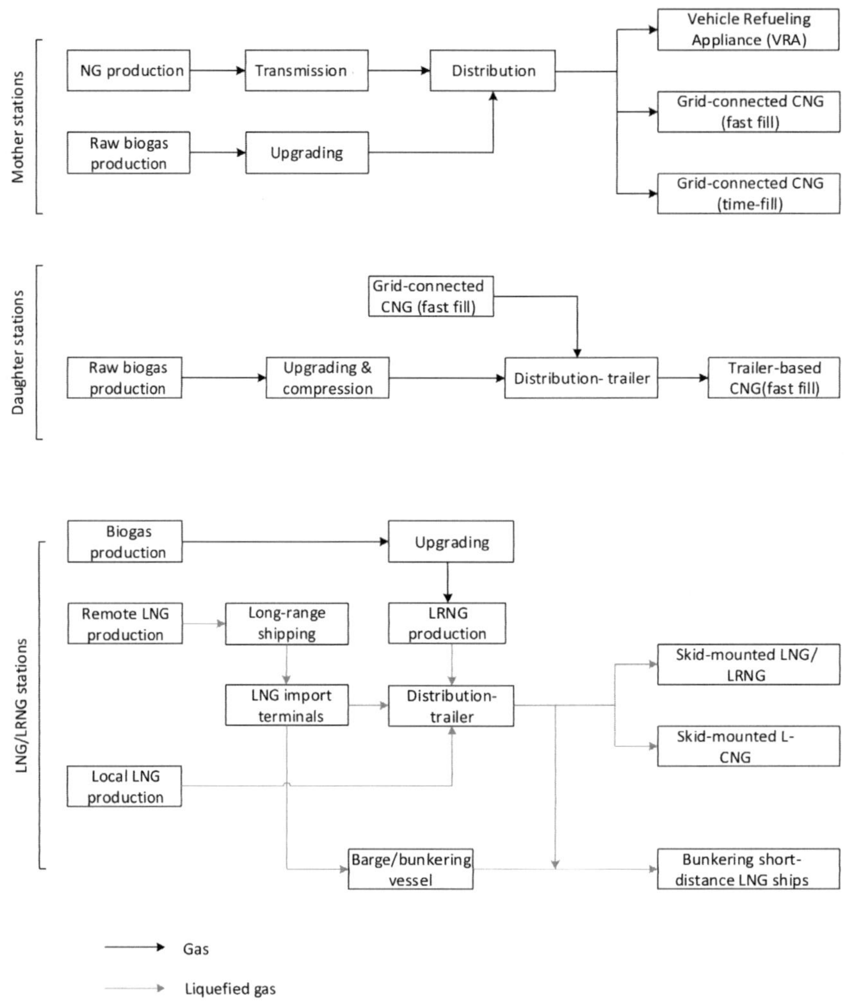

Figure 2. Schematic diagram of the processes and energy flows of potential gas supply pathways included in the WTT evaluation.

Then, based on figure 2, a total of 14 pathways have been selected for cost evaluation as shown in table 1 (with their detailed descriptions and acronyms).

The mode of gas distribution to filling stations and bunkering ports are classed into four: a low-pressure (4 bar) polyethylene pipeline, CNG swap body/truck (4500 Nm^3 capacity), LNG trailer ($60m^3$ capacity), and small-scale LNG carrier (7500 m^3 capacity). Given the current practises, the one-way distance between loading and unloading terminals is assumed to be 1km for pipeline, 50km for CNG/CRNG swap body/ truck, and 950km for LNG trailer and LNG carrier vessel. Also, the filling stations capacity assumed to be 1300 kg/day for all cases, except for home filling or private stations, which is assumed to be 36 kg/day.The daily vessel bunkering capacity assumed to be $10m^3$ per vessel. Pipeline cost functions (based on diameter, distance, and urban/rural share) [6], biogas plants investment costs [7, 8], LNG carrier/truck/trailer investment costs [6, 9], and filling station costs [10, 11] are taken from their respective sources and adjusted for 2016 price level using GDP deflators.

2.3 Gas vehicles included

The gas vehicles included in the evaluation are passenger (HDVs) for road-transport applications, and a short-range passenger vessel for maritime transport applications. The three state-of-the art gas engine technologies are port injection spark ignition (PISI), port injection dual-fuel (PIDF), and high pressure direct injection (HPDI) engines. The PISI gas engine works on 100% gas and Otto cycle, and it has a lower compression ratio and volumetric efficiency, and, thus, a lower cycle efficiency (about 35%), compared to diesel engine [12].

Table 1. Selected pathways and process description.

Type	Pathway acronym	Final fuel	Pathway description
Mother stations	VRA	CNG	Danish NG mix, distributed through transmission and distribution pipes to grid-connected households/industries. The home-filling facility, called vehicle refuelling appliance (VRA), is assumed to be supplied with a low-pressure grid (4 bar).
	CNGMF CNGMT	CNG	Danish NG mix, distributed through transmission and distribution pipes to grid-connected filling stations. The station could be either a fast-fill (CNGMF) or time-fill (CNGMT) station connected with a low-pressure grid.
	CRNGP-waste CRNGP-manure	CRNG	Raw biogas production from municipal organic waste (CRNGP-waste) and manure (CRNGP-manure), upgrading, and injection into the low-pressure grid (4 bar) through plastic pipes.
Daughter stations	CNGD	CNG	The same process description as CNGMF pathway, but it represents daughter stations. CNG supplied to the station is assumed to be filled at mother station and transported with truck/CNG trailer.
	CRNGD-waste CRNGD-manure	CRNG	Raw biogas production from municipal organic waste (CRNGD-waste) and manure (CRNGD-manure), upgrading and compression to 200 bar, and truck/CNG trailer distribution to fast-fill CNG filling station.
LNG/LRNG stations	LNG	LNG	Remote LNG production, LNG sea transport to north-western Europe import terminals, distribution by truck/LNG trailer to skid-mounted LNG filling stations.
	L-CNG	CNG/LNG	The same as LNG but at filling stations both LNG and CNG are available. Also, includes LNG vaporisation/compression to CNG at skid-mounted L-CNG.
	LNG-STS	LNG	Remote LNG production, LNG sea transport to north western Europe import terminals, distributed by LNG bunkering vessel to bunkering facility at ports (storage tank); ship-to-storage (STS).
	LNG-TTS	LNG	The same process description as LNG-STS, but LNG assumed to be distributed by truck/LNG trailer to bunkering facility at ports (storage tank); truck-to-storage (TTS).
	LRNG-waste	LRNG	Raw biogas production from waste, upgrading, liquefaction, and local distribution by truck/LNG trailer to LNG filling stations.
	LRNG-manure	LRNG	Raw biogas production from manure, upgrading, liquefaction, and local distribution by truck/LNG trailer to LNG filling stations.

The PIDF engine works on both diesel and gas; 50-60% gas (on energy basis) without compromising emissions. For maritime transport applications, due to the higher engine stroke to bore ratio, the share of gas could be as high as 95%. The engine is usually optimised for gas operation and exhibits similar efficiency to conventional diesel engine (about 43%) [12]. The HPDI gas engine works on the diesel cycle with 90-95% gas; a small amount of diesel is used to assist ignition to compromise NG's lower cetane number. As opposed to PISI and PIDF, where both CNG and LNG can be used, in a HPDI engine, only LNG is used, due to the required high pressure (about 300 bar) at the cylinder head. The assumed fuel economy for the medium size gasoline and CNG cars is 19 km/L and 3.74 kg/100 km, respectively. For diesel cars, it was adjusted with the assumed engine cycle efficiencies. For HD vehicles, 31 L/100km and 28 kg/100km (for PISI engine), respectively. Also, for conventional passenger vessel to be 0.02 km/L; estimated with a capacity of 600 passengers cruising at 14 knots.

3 Results and discussions

Based on the assumptions discussed briefly in Section 2, the WTT costs of each selected pathway and the break-even vehicle added costs for road transport and maritime transport applications have been investigated in detail, and the results are presented as follows.

3.1 WTT fuel production and conditioning costs

Figure 3 shows the fuel production and conditioning costs at each process stage, for each pathway. The gasoline and diesel fuel production costs are the average prices (excluding taxes) in Denmark in 2016, and used as a reference to compare the gas pathways.

 https://doi.org/10.1051/e3sconf/20185101008

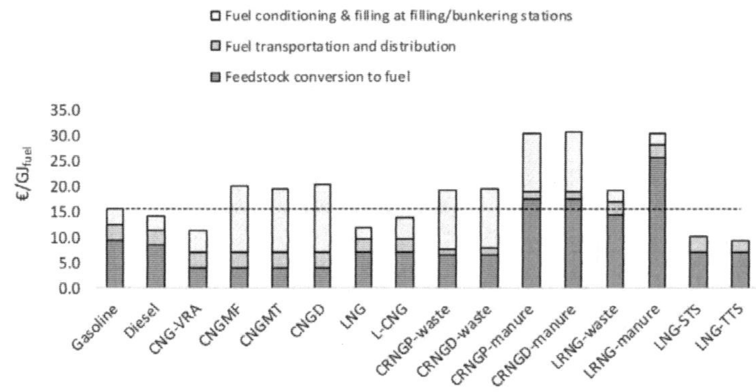

Figure 3. WTT fuel production and conditioning costs for all selected pathways. It also shows the cost break-down by process and activity.

As shown in figure 3, there exist substantial fuel production cost differences between CNG pathways, owing to the alternate gas transportation mechanism and filling station configuration: pipeline vs truck/trailer and fast-fill vs time-fill. CNG-VRA was found to be the cheapest pathway with an 11.2 €/GJ; however, all costs are to be transferred to private individuals, and affordability of the upfront investment cost is a challenge. Compared with the gasoline/diesel production cost, all CNG pathways, except CNG-VRA, showed marginal advantage. Despite its long-distance shipping (10,000km) and distribution (via the GATE terminal to filling stations in Denmark, 1800km round-trip distance), LNG/L-CNG pathways, on average, showed a 6.3 to 8.3 €/GJ unit cost advantage over CNG pathways.

Owing to the high investment and upgrading costs, both CRNG-manure and waste pathways showed a higher unit cost over their counterparts, diesel/gasoline and CNG pathways (26.1 and 26.8 €/GJ, respectively). The subsequent upgrading and liquefication investment costs increase the overall unit cost of LRNG-manure and waste pathways further by more than 20%. Comparing imported LNG with local liquefaction of biogas, the difference is quite substantial; it is not competitive over LNG, though it avoids the logistical problems in distribution and handling of the compressed biogas. It is worth nothing that, owing to the liquefaction cost of 8.14 €/GJ (calculated for LRNG pathway) and fuel production cost 7.2 €/GJ (calculated for CNGMF), small-scale LNG production would cost 15.34 €/GJ, which costs above imported LNG. Therefore, the choice between liquefying and not liquefying the upgraded gas relies on the trade-off between the added 13% production cost and logistical advantages of liquefying biogas (LRNG).

It is also worth mentioning that, on average, filling station costs make up 50% of the total cost for all CNG pathways, but only 30% for LNG/L-CNG pathways, indicating that the handling costs of LNG are much lower than for CNG.

In this study, bunkering costs at ports are assumed to be 'sunk' costs, as it is not the only activity going on at ports/harbours. Therefore, the unit cost of fuel supply excludes the bunkering cost. Thus, compared to road-transport filling stations', both LNG-TTS and LNG-STS pathways have shown a lower cost. Comparing between LNG-TTS and LNG-STS, with the assumed 1,800km, round-trip distance and a $10m^3$ daily demand per vessel, a trailer distribution was found to be the cheapest alterative over an LNG carrier, and hence, the LNG-TTS pathway showed a lower cost than LNG-STS

3.2 Break-even vehicle added investment cost

To analyse the trade-off between vehicle added cost and fuel price advantage or price gap, a lifecycle cost (LCC) comparison of gas vehicles with their corresponding counterparts, conventional vehicles, was made. It has three parts: (1) gasoline and diesel passenger cars vs PISI gas car; (2) HD diesel vehicle vs PISI, PIDF, and HPDI HD gas vehicles, and (3) LS-MGO passenger vessel vs PISI, PIDF, and HPDI HD LNG passenger vessels. The break-even added cost is the maximum vehicle added investment cost of a gas vehicle over its corresponding conventional vehicle at no-loss no-profit point, or zero NPV. For a given price gap and annual driving distance, the net benefit of a gas vehicle can be calculated by subtracting the market vehicle added cost from the break-even vehicle added cost. Also, the price gap indicates the price difference between diesel/gasoline and CNG/LNG on energy basis.

3.2.1 Passenger cars

Figure 4 shows the break-even added cost of gas cars over gasoline and diesel passenger cars. Evaluating the current performance of gas cars over gasoline cars, the break-even added cost at 0.33 €/L price gap (1.4 €/L

 https://doi.org/10.1051/e3sconf/20185101008

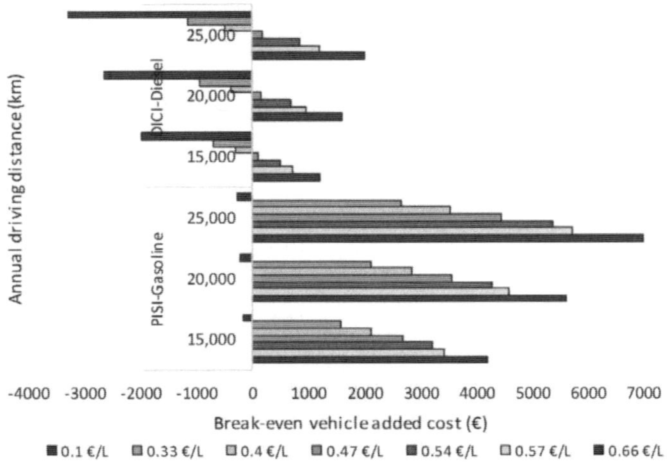

Figure 4. Break-even vehicle added cost of gas car for varied price gap and annual driving distance.

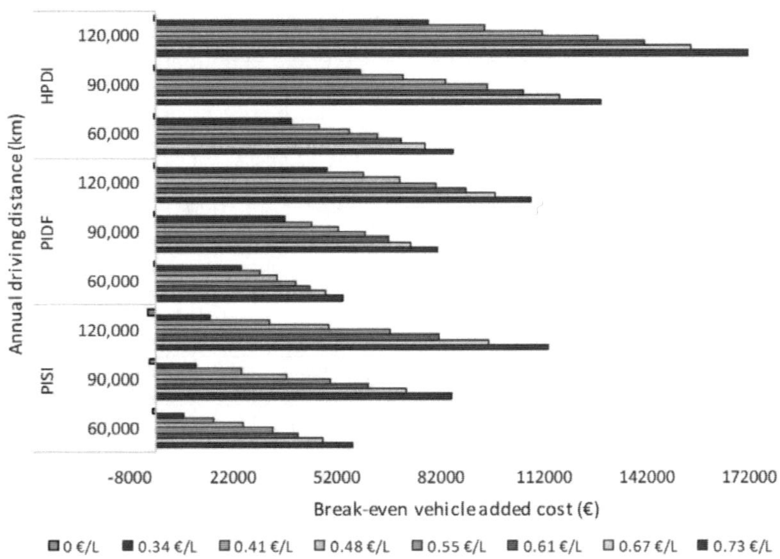

Figure 5. Break-even vehicle added cost of HD gas vehicle for varied price gap and annual driving distance. The price gap indicates the price advantage of LNG/CNG over diesel. The assumed current market prices at filling stations are: diesel (1.17 €/L), CNG (1.6 €/kg), and LNG (1.03 €/kg).

gasoline and 1.6 €/kg CNG price in Denmark) is estimated to be 1,585, 2,113, and 2,642 € for an annual driving distance of 15,000, 20,000, and 25000km, respectively. This means that, to be profitable, the current market added cost of gas cars should be less than the estimated break-even costs. For example, as of June 2017, the highest selling gas car in Sweden, the VW 1.4 TGI 110 BlueMotion, has an added cost of 1630 € [13]. In [11], the average added cost is reported to be between 700-1,050 €. The saving of VW 1.4 TGI 110 BlueMotion, in our case, would only be 483 and 1,012 € if the annual driving distance is 20,000 and 25,000km, respectively.

Since, in most cases, the average annual driving distance of a passenger car is below 15,000km – 11,037km in Denmark, 12,240km in Sweden, 12,387km in Norway – increasing the price gap is a more realistic way should gas cars be of interest. It is also important to note that the break-even added cost is marginally more sensitive to gasoline price than CNG price. For example, a 10% reduction in CNG price would increase the price gap to 0.44 €/L while a 10% increase in gasoline price would result in 0.47 €/L.

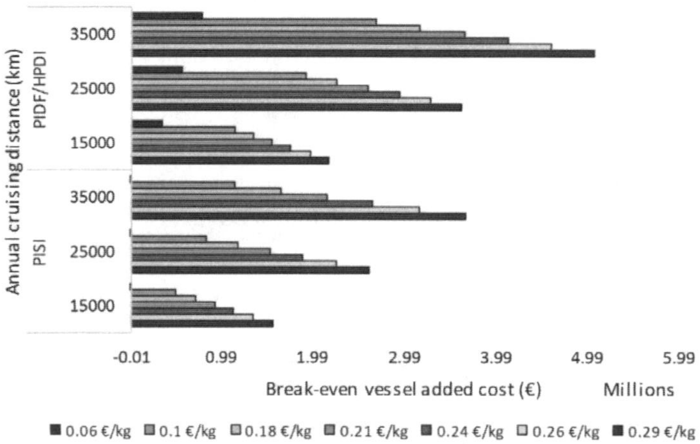

Figure 6. Break-even added cost of LNG passenger vessels for varied price gap and annual cruising distance. The price gap indicates the price advantage of LNG over LS-MGO, in terms of €/kg of LS-MGO. The assumed current market prices at ports are: LS-MGO (0.51 €/kg) and LNG (0.48 €/kg).

Compared to gasoline, diesel is cheaper, at 1.17 €/L, and with the current 1.6 €/kg CNG price, the price gap is zero. As seen in figure 4, there is a substantial difference between the gasoline and diesel car cases. Conventional diesel cars are more efficient than gasoline cars. The lower price gap and increased fuel economy contributes to the negative break-even added cost. Negative break-even added cost indicates a net loss. Even at a 0.66 €/L price gap (corresponding to a 1.83 €/L diesel price) the break-even added cost is only 1,200 € at 15,000km, while the cost increases to 2,000 € at 25,000km. Normally, for similar brake power output, diesel cars are more expensive than gasoline cars by about 2,000 € [13], and comparing a diesel with a gas car would be an even more expensive option.

3.2.2 Heavy duty vehicles

Figure 5 shows the comparison made for HD diesel and gas vehicles. Three cases are drawn for the HD gas vehicle: PISI, PIDF, and HPDI engine. CNG is the only fuel used in PISI while LNG is used in both PIDF and HPDI engines. Technically, both CNG/LNG could be used in a PIDF engine. The effect of higher fuel consumption and longer driving distance in HDVs on the break-even added cost can be noted by comparing figure 4 and figure 5. For the same price gap, the break-even added cost is significantly higher in the HDV application. Evaluating the current performance of the PISI engine HD vehicle, the price gap is zero (1.17 €/L and 1.6 €/kg), which results in negative break-even cost at all driving distances: -1,216, -1,823, and -2,430 € at 60,000, 90,000 and 120,000 km, respectively. On average, the added cost for HD gas vehicles is reported to be between 10,600-16,450 € [11]; assuming this, for PISI-engine profitability, a price gap higher than 0.34 €/L or 29% reduction in CNG price is required at 60,000km. LNG is much cheaper than CNG; the current price gap is about 0.42 €/L. Evaluating the current performance, the break-even cost is estimated to be 30,000 and 47580 € for PIDF and HPDI engine vehicles, respectively, at 60,000km. On average, for both engines, the break-even cost increases by 50% and 100% at 90,000km and 120,000km, respectively. Most refuse trucks and city buses travel less than 60,000km per annum while long-haul truck/trailers travel more than 100,000km. For example, in Sweden, the average annual driving distance of a bus is about 56,960km [14]. A HPDI gas engine vehicle would be, economically, more competitive in long-haul transport application owing to the higher diesel fuel substitution rate. The added cost of HPDI is normally higher than PIDF due to new accessories (like on-board high-pressure fuel pump), but the higher break-even cost could potentially outweigh the added cost benefit of PIDF.

A comparison of our results with prior studies shows that the results are in line with [15], though it shows the estimated break-even distance (annual traffic volume) instead of added vehicle cost (the other way round). For a 0.45 €/L price gap and 42,000-50,000 € added vehicle cost, the break-even distance was estimated to be 123,000, 360,000, and 456,000km for refuse trucks (CNG), transit buses (CNG), and long-haul fright trucks (LNG), respectively. The correlation between price gap and break-even distance was found to be a gently decreasing exponential function.

3.2.3 Passenger vessels

Figure 6 shows the break-even cost of LNG vessels cruising in Emission Control Areas; in relation with LS-MGO passenger vessel. Since, for maritime transport applications, the share of gas substitution in PIDF is above 95%, there is no distinction between the PIDF and HPDI cases here. The results indicate that a minimum of 0.1 €/kg is required to break-even at the assumed minimum annual cruising distance. For the assumed highest annual cruising distance (35,000 km) and highest price gap (0.29 €/kg), the break-even costs estimated to be as high as 5 M€ and 3.67 M€ for PIDF/HPDI and PISI LNG vessels, respectively. Potentially, the break-even cost shows the cost advantage between installing sulphur scrubber and swapping for high quality fuels like LNG. Owing to the higher fuel consumption of maritime transport per km travel, in general, for high frequency and short distance routes, as most ferries in Denmark, swapping the existing ferries for LNG is advantageous over LS-MGO.

4 Conclusions

In this study, 14 potential gas supply pathways were evaluated for road transport and maritime transport applications through WTT fuel production costs evaluation and break-even added investment costs of NGVs. The pathways are evaluated in the Danish gas (natural/biogas) context.

The results indicate that the gas distribution mode and filling stations configuration largely impact the WTT fuel production costs of the selected gas pathways; in most cases, filling station costs alone make up 50% of the total costs in all CNG pathways and 30% in all LNG/L-CNG pathways. The large economy of scale outweighs the long-distance shipping and distribution costs of imported LNG, and showed the least-cost of all gas and diesel/gasoline pathways. Due to the high investment costs, the choice between liquefying and not liquefying the upgraded biogas relies on the trade-offs between the added production costs and logistical advantages of liquefying biogas. The price gaps and annual driving distances are critical for gas vehicles cost competitiveness. For a given price gap, the higher the annual fuel consumption, the higher would be the annual savings and the break-even added investment costs of gas vehicles. Lastly, the results, however, are very sensitive to the various critical assumptions that made throughout this paper, such as, the investment costs, gas distribution distances, and others. Hence, the results should be interpreted with that in mind.

Acknowledgments

The authors are grateful for the financial support from the Innovation Fund Denmark under the FutureGas project.

References

1. European Enviromental Agency. *Transport greenhouse gas emissions*. 2017 [cited 2017 February 16]; Available from: http://www.eea.europa.eu/airs/2016/resource-efficiency-and-low-carbon-economy/transport-ghg-emissions.
2. International Association for Natural Gas Vehicles (IANGV). *Latest International NGV Statistics*. 2016; Available from: http://www.iangv.org/.
3. NGV Global. *Natural Gas Vehicles by country*. 2016 [cited 2016 December 22]; Available from: http://www.iangv.org/stats/NGV_Global_Stats1.htm.
4. International Maritime Organization (IMO). *Sulphur oxides (SOx) – Regulation 14*. 2016 [cited 2017 January 04]; Available from: http://www.imo.org/en/OurWork/Environment/PollutionPrevention/AirPollution/Pages/Sulphur-oxides-(SOx)-%E2%80%93-Regulation-14.aspx.
5. Eise Fokkema, J., P. Buijs, and I.F.A. Vis, *An investment appraisal method to compare LNG-fueled and conventional vessels*. Transportation Research Part D: Transport and Environment, 2017. **56**: p. 229-240.
6. H2A Analysis Group. *H2A Delivery Components Model version 2.0*. 2017 [cited 2017 May 29]; Available from: https://www.hydrogen.energy.gov/h2a_analysis.html.
7. ENERGINET.DK, *Technology data for energy plants*. 2012: Denmark. p. 157.
8. Bauer, F., et al., *Biogas upgrading – Review of commercial technologies*. 2013, Swedish Gas Centre (SGC).
9. Stolt-Nielsen Limited. 2017 [cited 2017 August 12]; Available from: https://www.stolt-nielsen.com/en/press-release/?id=566.
10. Smith, M. and J. Gonzales, *Costs Associated With Compressed Natural Gas Vehicle Fueling Infrastructure*. 2014: Denver, USA. p. 15.
11. Kollamthodi, S., et al., *The role of natural gas and biomethane in the transport sector* 2016, Ricardo Energy & Environment United Kingdom
12. Hegab, A., A. La Rocca, and P. Shayler, *Towards keeping diesel fuel supply and demand in balance: Dual-fuelling of diesel engines with natural gas*. Renewable and Sustainable Energy Reviews, 2017. **70**(Supplement C): p. 666-697.
13. Volkswagen (VW) Sweden. *Price list*. 2017 [cited 2017 August 30]; Available from: https://www.volkswagen.se/sv/broschyrer.html.
14. SIKA, *Values and methods of the transport sector socioeconomic analysis*. 2009: Östersund, Sweden. p. 122.
15. Cai, H., et al., *Wells to wheels: Environmental implications of natural gas as a transportation fuel*. Energy Policy, 2017. **109**: p. 565-578.

E3S Web of Conferences **51**, 01009 (2018) https://doi.org/10.1051/e3sconf/20185101009
ICACER 2018

Analysis-based results on the delineation of prearrangement areas for marine renewable energy installations in the Western Black Sea Basin

Razvan Mateescu[1,3,*], Elena Vlasceanu[1,2] and Liliana Rusu[3]

[1] National Institute for Marine Research and Development "Grigore Antipa", 300 Mamaia Blvd., RO-900581, Constanta, Romania
[2] Constanta Maritime University, 104 Mircea cel Batran Street, Constanta, Romania
[3] 'Dunarea de Jos' University of Galati, Department of Mechanical Engineering, 111 Domneasca Street, Galati, Romania

Abstract. The selection of optimum places for installations and suitable solutions for the use of marine renewable energy requires a complex approach as well as a longer period of time and a spatial scale hydrodynamic characterization, which time-limited and local studies are incapable to provide. To that end, the complex monitoring and modelling approach on the evolution of large-scale marine and coastal processes in response to both direct and indirect human intervention and to natural phenomena variability is an increasingly important issue in the Integrated Coastal Zone Management (ICZM) and MSP implementation, all over the world. The installation of wave/currents devices within coastal and marine areas involves the knowledge of the physical, economic and social impact of the continuous modification of the marine hydrodynamics, due to the new changes induced in the marine climate. The present work adequately takes into account the combined results of Marine Spatial Planning Directive applications within Western Black Sea Basin (WBSB). The MSP approach allows the selection of efficient areas for the installation of marine waves and currents devices converters (WECs), in relation with the potential environmental impact of adjacent MPAs, taking into account the induced strong variability of renewable devices/convertors on the coastal processes. Recommended areas for the installation of the above-mentioned devices within highly energetic areas asked for a proper response in order to develop a detailed delineation targeting a sustainable coastal protection.

1 Introduction

The coastal zone is constantly changing in ways that impact every facet of our society. To that end, it is well-known today that the coastal zones are among the most dynamic parts of the Earth. The systematic analysis, modelling/forecast and planning of these areas pose ongoing and very important challenges. This is important for the selection of the optimum solutions for socio-economic activities dealing with the marine environment, especially for coastal protection against the implementation of the technical solutions, in relation with renewable energy [1-3], and, last but not least, with respect to the enhancement of ecosystem-based management.

For Romanian touristic littoral, the waves' regime was considered based on monitoring and numerical modelling, in relation with new protection measures for coastal navigation [4], and also, for new balneary developments, mainly for the delineations of suitable sites for thalassotherapy, close by high energetic coastal areas [5].

Also, the deployment of marine energy conversion equipment in key areas of waves/currents intensification is mainly approached by certain case studies on the implementation of various technologies as well as on its environmental impact.

The methodology that was used to collect, process and transmit the energy of the marine environment is under the attention of different regional stakeholders, represented by public authorities, marine planners and regulators, research institutes, university and specific developers. Since the Black Sea is a European marginal sea, where strong storms are distinctive, the investigations on such energy developments are extended in specific sites, taking in account its cumulative impacts on the marine/coastal ecosystems.

Due to the north-western Black Sea (NWBS) shore configuration and the general characteristics of the environment, Black Sea waves and circulation processes have a dynamic which is different than that in the case of the ocean [2, 3]. Moreover, wind-generated waves in relation with inland water inflow and atmospheric phenomena induce specific hydrodynamic singularities in the coastal zones, noticed in the coastal system response modelling as well in case of marine energy usage.

* Corresponding author: razvan_doru@yahoo.com

Consequently, certain necessities of existent vulnerabilities connected with regional network of Marine Protected Areas (MPAs) or different marine activities are considerate by MSP institutions for future larger scale installations at proper locations of the NW Black Sea Basin/Shore.

The implementation of the above-mentioned devices requires a multidisciplinary approach on the general trend and specific induced phenomena on the NWBS littoral/sea-land-interface, towards the understanding and mitigation of the impacted marine/coastal processes and certain specific components of marine life. The identification and quantification of the changes/impacts upon the Romanian coastal zone environment of the NW Black Sea is viewed in relation with the possible harmonization of different marine/coastal activities.

2 Methodology

The common methodology applied is connected to Geographic Information System (GIS) analysis and maritime spatial planning. Collecting information describing the plan area - indicators that are relevant for the processes present in the administrative units and data (texts and maps) from recent documents [6], useful for the plan preparation, marine and coastal context data relevant for the influences on the plan area [7], spatial planning documents and strategies relevant for the areas of technology implementation.

The ensuing result of this approach are the plan areas that are geographically delimited, significant spatial data and information on external influences on the plan area, together with provisions of recent spatial development plans and strategies pertaining to the marine energy convertors disposal [8-11], legislation and regulations relevant for the implementation of EU documents targeting maritime spatial planning [12].

The current situation analysis focused on the ecological and anthropic processes present in the marine area and the factors that affect their sustainable existence, in relation with the influences of energy devices [13, 14].

The following domains were subjects of analysis:

- Natural environment features/protection (physical, biological and valuable natural features, biodiversity and habitats, MPAs);
- Marine and coastal spatial features and resources (marine/coastal habitats, marine living and non-living resources);
- General factors influencing the ecosystems (geographical and physical conditions of the marine and coastal areas, atmospheric/marine climate);
- Natural and anthropic harmful processes (quality status of the environmental constituents, natural and technological risks, pressures on marine environment);
- Built environment/Settlements network - coastal settlements affecting the marine space of possible implementations;
- Technical infrastructures networks (transport infrastructure, water management works, energy generation and transport, telecommunications, waste management);
- Marine and coastal activities/zoning - actual situation of marine habitats, marine areas for human uses (transport, fisheries, aquaculture, etc.), coastal areas (built-up areas, areas for agriculture, forests, protected areas, etc.), military training areas and facilities.

Analysis considered the NWBS supra-territorial context at regional level, such as transport corridors, development poles, and neighboring areas of ecosystems and the use of marine resources.

The spatial analysis of present (2017) in contrast to historic (1962) shoreline changes in relation with shore vulnerability index was realized in GIS environment.

3 Results and discussions

The results of this study provide an important support for the coastal planning/designing decisional process regarding the WQ/coastal ecosystem protection, erosion/accretion phenomena, the renewable energy/coastal facilities extensions.

The actual erosion state was emphasized on the base of shoreline retreat (figure 1) calculation in ArcGIS. The calculated results show that more than 75% of the total extent of Romanian shore is affected by erosion, therefore situation of the deployment of the renewable energy installations should be considerate as potential, if not compulsory, protection measures.

The key results provide an open perspective on the impact of the projects for marine energy, as well as on their relevance and direct application, jointly with a better knowledge of such results, potentially operational within wave and current climate in the NW Black Sea basin, including levels of impact on coastal dynamics. Certain realistic estimates on the installation of marine energy devices were identified at the mouths of the Danube River and the vicinity of the major marine obstacle/ports and natural capes/promontories of the NW BS Shore, in relation with environment ecological characteristics, Marine Protected Areas and marine/coastal activities and use [15]. Accordingly, facilitation, identification and selection of proper installation sites is possible, in keeping with the assessed historic shorelines and indicators of climate change vulnerability, and also in line with the relevant existing legislation and initiatives (Marine Strategy Framework Directive, Maritime Spatial Planning, and so on.

Environmental data analysis of existing conditions, existing economic sectors/uses and natural valuable areas was overlapped on a general layout highlighting the spatial representation of possible installation areas among identified marine activities and use of resources in the broader context of regional environmental conditions.

The main activities present on Romanian shore general layout (Figure 2) in relation with MSP approaches are shown together with MPAs (SCI network afferent to Constanta and Tulcea administrative units). The input of the RECZ areas and its specific considerations for Romanian maritime general layout was enquired by

https://doi.org/10.1051/e3sconf/20185101009

specific studies for maritime activities' harmonization and permit the completion of the MSP Directive implementation's processes [16].

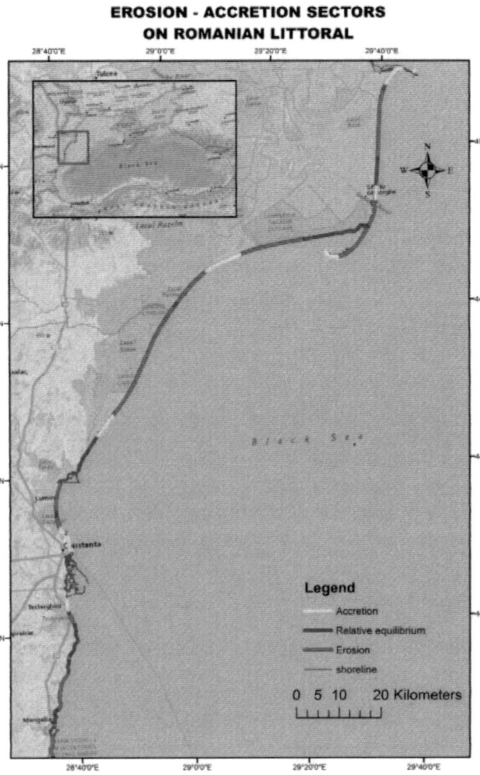

Figure 1. Coastal vulnerability sectors on Romanian littoral

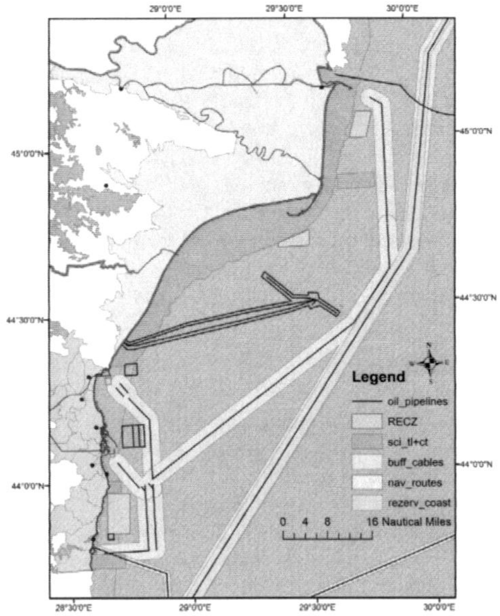

Figure 2. Map of major maritime activities/usage, marine fisheries areas, Marine Protected Areas and areas for installations of renewable energy convertors, according with MSFD implementation requirements on the NW Black Sea Basin.

4 Conclusions

The principles used in the present paper were the ecosystem - based approach (environment preservation, protection, improvement and resilience to climate change), sustainable development and growth in the maritime sector, and the harmonization the relevant maritime activities and levels of usage.

The present study is structured as a prequel to the prospective diagnosis approach and the extension of the implementation strategy of these technologies.

Acknowledgments

This work was supported by a grant of Romanian Ministry of Research and Innovation, CNCS – UEFISCDI, project number PN-III-P4-ID-PCE-2016-0028, within PNCDI III.

References

1. E. Rusu, L. Rusu, C.G. Soares, 9th International Workshop on Wave Hindcasting and Forecasting (2006)
2. E. Rusu, J. Mar. Sci. Tech. **14**, 359–372 (2009)
3. E. Rusu, J. Mar. Sci. Tech. **15 (2)**, 143–159 (2010)
4. L. Rusu, C.G. Soares, J. Mar. Sci. Tech. **19(2)**, 215-227 (2014)
5. A. Ivan, E. Vlasceanu, SGEM **17**, 561-569 (2017)
6. C. Gasparotti, E. Rusu, J. Environ. Prot. Ecol. **13** (3-A), 1751-1759, (2012)
7. E. Rusu, Energies **7 (6)**, 4002-4018, (2014)
8. E. Rusu, F. Onea, Renewable Energy **85**, 687-703 (2016)
9. S. Diaconu, E. Rusu, Sci. World J, ID 498013 (2013)
10. D. Niculescu, E. Rusu, Mechanical Testing and Diagnosis **2**, 25-34 (2016)
11. A. Ivan, C. Gasparotti, E. Rusu, J. Environ. Prot. Ecol. **13**, 1673-1682 (2012)
12. I. Omer, R. Mateescu, L. Rusu, D. Niculescu, E. Vlăsceanu, J. Environ. Prot. Ecol. **16(2)**, 417-424 (2015)
13. L. Rusu, Energies **8(9)**, 10370-10388 (2015)
14. E. Rusu, S. Măcuță, J. Environ. Prot. Ecol. **8(1)**, 147-151 (2009)

https://doi.org/10.1051/e3sconf/20185101009

15. L. Rusu, D. Butunoiu, E. Rusu, J. Environ. Prot. Ecol. **15(2)**, 445-454 (2014)
16. E. Vlasceanu, D. Niculescu, I. Omer, E. Rusu, A. Ivan, SGEM **3(2),** 675-682 (2015)

Solar Energy as Expedient Alternatives for Nuclear Energy

T Younas[a*], H Rasheed[b], M M Rehman[b], R Ramzan[b]

[c] Assistant Professor, Szabist, Karachi and 75600, Pakistan
[b] Student, Szabist, Karachi and 75600, Pakistan

Abstract. The development of solar power technologies is considered to be the key solutions for fulfilling ever increasing worldwide demand for energy. This paper reflects the nature, characteristics, similarities, and differences between nuclear and solar energies. It also discusses their technologies, their impact on different aspects of life and public attitudes about them. Generally, nuclear energy is considered to be destroyed due to uncontrollable chemical reactions associated with harvesting of nuclear energy. Energy requirement of the world is increasing day by day, and using conventional methods to fulfil this requirement might cause devastating effects on atmosphere due to an emission of different harmful by-products released. In this paper, alternate methods of producing energy are discussed and a detailed analysis of its environmental effects, sustainability and cost have been discussed.

1 Introduction

An Energy crisis is one of the big problems of the world nowadays. There are as many as ten main resources of energy which are being used in the different part of the world. We would be discussing and comparing only two of them i-e Nuclear energy and Solar energy. First, we need to understand what nuclear energy is. According to Einstein's equation it is described below:

$$E = mc^2 \quad (1)$$

Mass (m) can be converted into energy (E) in nuclear process. In the equation, c is speed of light. Therefore, the more mass has we used the more energy we produce [1]. We cannot see an atom and its sub-particles with our open eyes because they are too small, so their counting is not possible. However, scientists (particularly physicists and chemists) have been studying atoms for over 200 years and have developed models to explain the results of their experiments.

In our solar system, the sun is the only star which is stationary and all planets including the Earth are rotating around it. All life on Earth is supported by the energy in the form of solar radiation gathered from the sun. It is done through the system known as photosynthesis. It also controls the climate and weather of the Earth [2]. The energy requirements of the entire world for a full year can be met by the sunlight coming to Earth in an hour and a half [3]. Solar energy is the oldest energy source ever used. Solar power is an ideal power source we can capture with solar panels and other collectors. Once these collectors are manufactured, there is no pollution or waste product. At the same time, there are no moving parts to hurt wildlife. Solar energy has the potential to full fill the energy demands of the entire world if technologies for its harvesting and supplies or made available [4]. Adoption of solar technologies would be mitigated all issues connected with energy security, climate change, and unemployment. It will also play an important role in transportation sector because it does not require any fuel transportation. The annual solar distribution over the Earth surface is shown in figure 1.

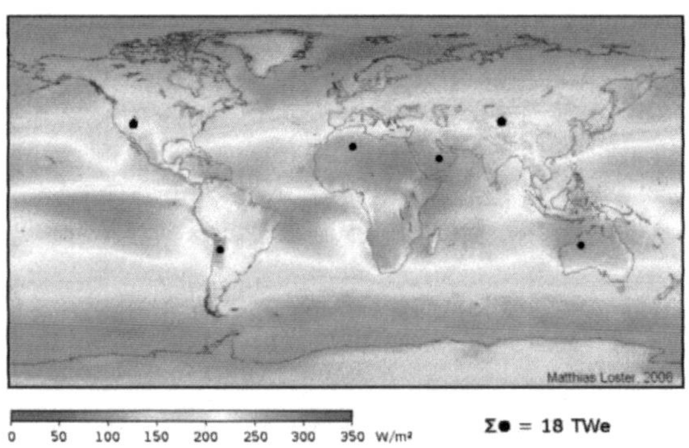

Figure 1. Annual average solar distribution [5].

2 Comparison between nuclear and solar energies

2.1 Status of technologies

Nuclear power generation system comprises the nuclear reactor and its energy conversion system. To achieve the aims of generation system, four areas are required to develop and these are:

* Corresponding author: Tanzila@szabist.edu.pk

 https://doi.org/10.1051/e3sconf/20185102001

1. Sustainability
2. Safety and reliability
3. Proliferation resistance
4. Physical protection and economies

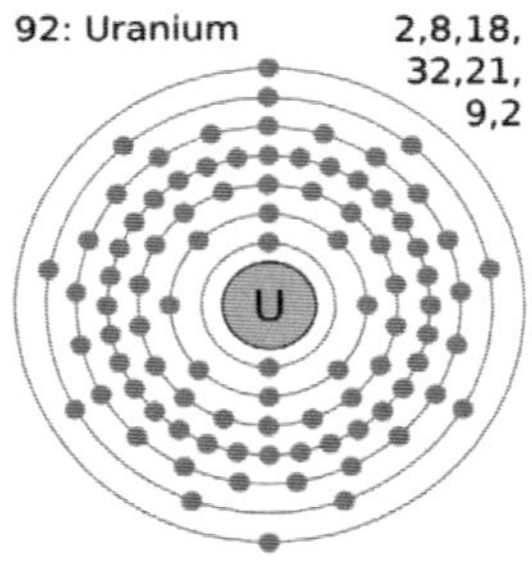

Figure 2. Uranium Model [6].

Uranium model shown in figure 2 tells us that nuclei are continuously breaking up and throwing out neutrons. Experiments show that if a neutron hits or is absorbed in the nucleus of another uranium atom, the nucleus may split into two parts and releases two or three neutrons and give off energy. This process is called "fission" and the energy is called "nuclear" energy. The fission reaction is shown in figure 3

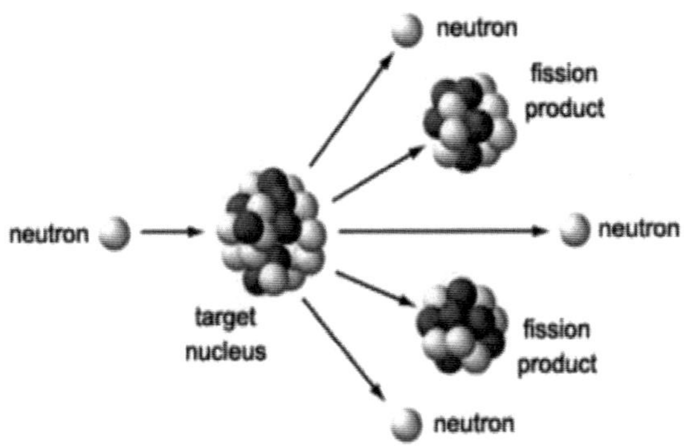

Figure 3. Fission Reaction [7].

An active solar system collects solar radiation and uses mechanical and electrical equipment for conversion of solar energy to heat and electric power. The most popular application of the system is solar water heater system.

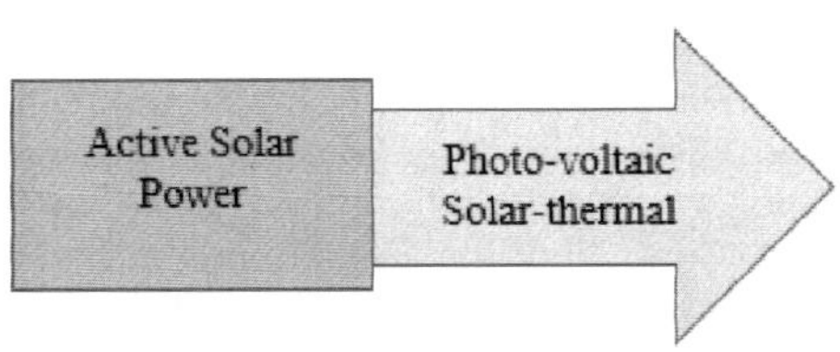

Figure 4. Classification of Active Solar Power.

There are two kinds of active solar energy technology as shown in figure 4. One is photo-voltaic technology, and the other is thermal technology. In photo-voltaic technology semiconductors are used to convert sunlight directly into electric energy. Whereas in thermal technology, solar energy is harnessed into thermal energy for domestic as well as commercial application such as drying, heating, cooling, cooking, etc. Concentrating solar power (CSP) is typically used electricity plants and is not beneficial for the residential purpose. This technology makes use of mirrors to reflect and concentrate sunlight onto receivers that collect solar energy and convert it into warmth that could later be used to provide electricity [8]. The critical raw material of solar energy technologies is shown in the table 1.

Table 1. Critical Raw Materials Content. [9].

Application	Components	Critical raw materials
Photo-voltaic and Thermal	Semi-conductor	Cadmium, Indium, Gallium, Germanium Selenium, Tellurium
Concentrating solar power (CSP)	Electric contacts mirror	Silver

2.2 Public Support

It is very interesting to note the public opinion differs from time to time for support or rejection of any energy sources. Following table shows that as much as 90% public support is in favour of solar power energy. Whereas 38% support is enjoyed by nuclear energy. The public support for various energies is shown in the table 2 and figure 5.

Table 2. Public support in percentage for various energies [10].

Energy resources	Public support (%)
Solar power	97
Wind power	93
Hydroelectric power	91
Natural gas	80
Coal	48
Nuclear energy	38

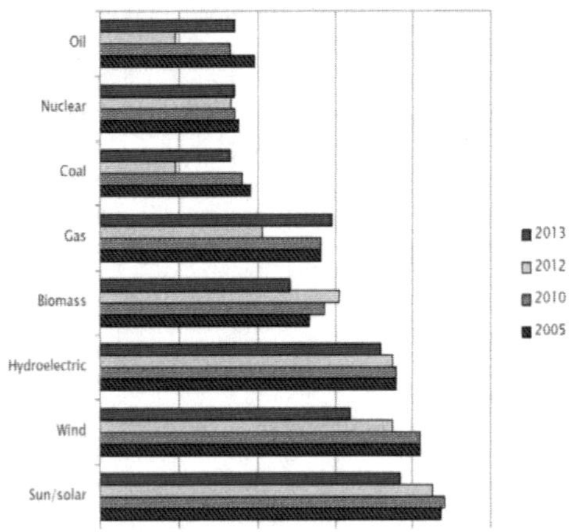

Figure 5. Changing attitudes to popular support for various technologies [11].

2.3 Accidents

In this section, we will describe serious incident and accidents at nuclear power stations. Since first of its kind occur in 1952in Canada [12]. We have identified 33 incidents and accidents from 1952 to date. In fact, international atomic energy authority does not have a complete historical database of such incidents and accidents. Out of 33, six had happened in the US, five in Japan and UK and Russia have had three. International atomic energy authority ranks these incidents and accidents using a special international nuclear scale ranging from anomaly to major accidents and has numbered them from one to seven. There has been only one scale seven incident in history at Chernobyl in 1886 [13]. The events at Fukushima were level five only. Placing solar panels on roof is quite dangerous only in Australia about 30 people died every year due to falling from height during installing solar panels. However solar panel may ultimately be safer than other energy resources because of the reduction of pollution and putting solar panel on ground.

3 Application of solar energy

Considering the usage of energy and energy requirements, a lot of people are considering it to be the alternate energy source. Considering eco-friendly nature of solar energy, solar energy has a great number of applications. These applications have a wide spectrum from industrial to domestic use. From wide range, some of the applications are discussed in this paper [14].

3.1 Solar water heating

Solar water heating unit consist of blackened flat plate metal collector with an associated metal tubing generally facing toward the sun. A layer of transparent glass and insulation is placed between metal tubing.

3.2 Solar distillation

In remote areas, clean water is one of the major issues. A solar distillation plant can be installed in remote and arid areas and this issues can be resolved. This solar distillation process works on the principle of absorption of solar radiation. These radiations are converted into heat and through this heat water is evaporated. This evaporated water is potable and is collected in a collection unit of the plant.

3.3 Solar furnaces

Solar furnaces can be used for processes that require high temperature. It uses a number of rotate-able mirrors which can be used to focus the solar beams. By adjusting mirrors, solar beams can be focused at a single point and can cause a high temperature

3.4 Solar electric power production

In Photo-voltaic cells are used for direct production of electricity. These photo voltaic cells are made of semi-conductors releases electrons when photons strike on the photo voltaic cells. These cells are arranged in series or parallel combination to form modules. These electric power stations vary in sizes according to the usage. These units can be used in powering up street lights as well as charging batteries in satellites

3.5 Solar green house

Solar greenhouse consists of large transparent structures that acts as a solar cell. Energy collected through this process can be used to produce heat for growing plants. As the radiations trapped in the house cannot escape, air inside the greenhouse gets enriched with CO2 which helps in better growth of plants.

4 Discussion

Nuclear power plants rely on uranium mining or processing which makes it non-renewable energy whereas renewable solar power generate utilizes sunlight which makes it renewable one. Nuclear energy does not emit greenhouse gasses directly but nuclear fuel cycle produces them indirectly. As solar energy does not contain any element, it is considered clean and environmentally friendly source of energy. Nuclear power is a high risk because of the chemicals in it. It can destroy the whole earth due to the radiation in it. Nuclear energy installation process is not very common so this type of energy is not available to everyone. Whereas solar energy simple in nature can be available to everyone.

5 Conclusion

We can easily conclude that solar energy does not only need of the hour but also need of the future because of its pollution-free nature, safety, and convenience. Solar energy has wider usage in all fields of society and is one of the energies consider as infinite for humankind. Solar energy is available even outside the solar system whereas it is not so with nuclear energy. We can easily say that solar energy will develop well and become far more pollution-free than now. Both these energies have low greenhouse gas emissions and high-cost electricity. They still differ in many ways like safety and principle of operation of devices It is obvious that the application of nuclear and solar energy will grow in future but solar energy is more promising.

References

1. The physics of the universe. [Online]. Available: http://www.physicsoftheuniverse.com/topics_relativity_emc2.html
2. Space based solar power [Online]. Available : https://en.wikipedia.org/wiki/Space-based_solar_power
3. Solar energy technology basis, August 16, 2013 [Online]. Available: https://energy.gov/eere/energybasics/articles/solar-energy-technology-basics
4. Conserve Energy Future [Online]. Available: https://www.conserve-energy-future.com/alternativeenergysources.php
5. Renewable and nuclear electricity: Comparison of environmental impacts Energy Policy, Volume 96, September 2016, Pages 758-769 by Charles McCombie, Michael Jefferson
6. [Uranium model [Online] Available: https://www.quora.com/Is-uranium-%E2%80%9C-92-%E2%80%9D-possible-uranium-nucleus-without-electrons-e-g-through-ionisation
7. Nuclear fission [Online] Available: http://www.atomicarchive.com/Fission/Fission1.shtml
8. Concentrating Solar Power Basics, August 20, 2013 [Online] Available : https://www.energy.gov/eere/solar/articles/concentrating-solar-power-basics
9. Edenhofer, O., et al., 2011. Renewable Energy Sources and Climate Change Mitigation: Special Report of the Intergovernmental Panel on Climate Change. Cambridge University Press, Cambridge
10. Wallard, H., Duffy, B, Cornick, P., 2012. After Fukushimal, Global Opinion on Energy Policy, IPSOS Social
11. Solar energy: Potential and future prospects Renewable and Sustainable Energy Reviews, Volume 82, Part 1, February 2018, Pages 894-900 by Ehsanul Kabir, Pawan Kumar, Sandeep Kumar, Adedeji A. Adelodun, Ki-Hyun Kim
12. Nuclear power plant accidents [Online] Available: https://www.theguardian.com/news/datablog/2011/mar/14/nuclear-power-plant-accidents-list-rank
13. Green deaths: The forgotten dangers of solar panels by Gavin Atkins, 17 May 2011 [Online] Available: https://asiancorrespondent.com/2011/05/green-deaths-the-forgotten-dangers-of-solar-panels
14. Solar Energy: Major Application of Solar Energy Article shared by : Puja Mondal [Online] Available: http://www.yourarticlelibrary.com/energy/solar-energy-10-major-application-of-solar-energy-explained/28197

Weather data errors analysis in solar power stations generation forecasting

Stanislav Eroshenko[1] and Alexandra Khalyasmaa[1*]

[1]Ural Federal University named after the first President of Russia B.N. Yeltsin, Ekaterinburg, 620002 Mira str. 19, Russia

Abstract. The paper presents a short-term forecasting model for solar power stations (SPS) generation developed by the authors. This model is based on weather data and built into the existing software product as a separate short-term forecasting module for the SPS generation. The main problems associated with forecasting the SPS generation on cloudy days were revealed in the framework of authors' research, which is due not to the error of the developed model but to the use of the same learning sample for both solar and cloudy days. This paper contains analysis of the main problems related to the learning sampling, samples pattern, quality and representativeness for forecasting the SPS generation on cloudy days. Besides, the paper includes a calculation example performed for the existing SPS and a detailed analysis of the forecast generation on cloudy days based on the actual weather provider data.

1 Introduction

Today, the main problem for the short-term (day ahead) forecasting of solar power stations (SPS) generation is the necessity to use an accurate and detailed weather forecast, which in turn is not cheap for the customer (SPS owner). In order to save money, the customer often uses the available free weather data, which, according to authors' research, is ultimately insufficient for accurate identifying the weather conditions necessary to solve the presented problem.

It should be noted that the weather data composition is of statistical significance from the point of view of the influence on the transparency factor calculation and allows to establish a reliable statistical relationship between the properties (cloud cover, the altitude of the solar disk) and the response (the solar radiation flux density on a horizontal surface) [1]. The impossibility of identifying the weather conditions leads to a high variance of the response y (transparency factor) with respect to the features under consideration xi (the estimate of cloud cover, the sine of the solar disk altitude angle).

An estimated evaluation shows that, on average, every fifth change in the actual value of the transparency factor does not correspond to a change in the actual cloud cover. At the same time, if in addition to take the forecast cloud cover into account, which is a priori less accurate than the actual one, the error of the initial data increases which introduces a significant error in the calculation results. To increase the accuracy of forecasting on days with precipitation, it is necessary to accumulate appropriate retrospective data and get a separate sample for days with precipitation. It is necessary to clearly distinguish the type of precipitation: drizzle, rain, rain shower, hail, snow, sleet, etc.

2 Short-term SPS forecasting model

The presented methodology is based on calculation of the solar radiation flux density (external illumination) on a horizontal surface for a six-day interval according to the weather provider data for half-hour intervals that is necessary for the regression model development. The methodology is described in detail in [2]. The main stages of the developed methodology implementation are presented below.

2.1 Forecast calculation of the solar radiation flux density incident on a horizontal surface`

The average value of the solar radiation flux density near the ground, incident on the horizontal surface, $\overline{G}$ for each hour of the forecast day is determined by the expression:

$$\overline{G} = \overline{k}_T \cdot \overline{G}_0, \tag{1}$$

where $\overline{G}$ is the average value of the solar power flux density near the ground for each hour of the forecast day, [W/m^2]; $\overline{G}$ is the transparency factor for each hour of the forecast day, [p.u.]; $\overline{G} = \overline{k}_T \cdot \overline{G}_0$ is the average value of the solar radiation flux density at the boundary of the atmosphere for each hour of the forecast day, [W/m^2].

To define a parameter $\overline{G}$ a physical model is used that allows to determine solar power flux density at the boundary of the atmosphere, while only the location of the power station, the number of the day in question and the hour of the day in question are needed. The algorithm

* Corresponding author: stas_ersh@mail.ru

for calculating solar power flux density at the boundary of the atmosphere is described by the following formulas.

- The average value of the solar radiation flux density at the boundary of the atmosphere $\overline{G}_0$ [3]:

$$\overline{G}_0 = \frac{12}{\pi} G_{on}(\cos\varphi \cdot \cos\delta \cdot (\sin\omega_2 - \sin\omega_1) + \frac{\pi(\omega_2 - \omega_1)}{180^0} \cdot \sin\varphi \cdot \sin\delta) \quad (2)$$

To define a parameter k_T a statistical model is used that allows determining the transparency factor for the next day. Retrospective cloud data from previous days, retrospective data on the solar radiation flux density from previous days, forecast cloud data for the next day, data on location of the power station, the number of the day in question, and the hour of the day are needed.

The regression model is described by the following expression:

$$\overline{k_T} = a_1 + b_1 \cdot (\overline{cc})^2 \cdot \sin\overline{\alpha} + c_1 \cdot (\overline{cc})^2 + d_1 \cdot \sin\overline{\alpha}, \quad (3)$$

where $\overline{k_T}$ is the transparency factor, [p.u.]; $\overline{cc}$ is the cloud cover, [p.u.]; $\overline{\alpha}$ is the average value of the solar altitude angle within the defined time step, [degrees]; a_1, b_1, c_1, d_1 are coefficients determined empirically by regression analysis equations, [dimensionless values].

In order to minimize the manual labor of the operational staff of the SPS in all subsequent calculations, it was decided not to implement the estimate of cloud cover cc in its original form, since this requires a complex analysis of retrospective data of considerable depth (up to two years) for the cloud cover TCC and density of the solar radiation flux incident on the horizontal surface of the earth $\overline{G}$ in order to determine the effect of the cloud character on the magnitude $\overline{G}$.

In further calculations, the value TCC is equal to the total cloud cover TCC in relative units [4]:

$$cc = \frac{TCC}{100}, \quad (4)$$

where cc is the cloud cover, [p.u.]; TCC is the total cloud cover, [%].

The transparency factor $\overline{k_T}$ for the previous days is determined by the known values of the solar radiation flux density for the previous days $\overline{G}$, obtained from the data of the local weather station, and the calculated values $\overline{G}_0$ according to the formula:

$$\overline{G} = \overline{k}_T \cdot \overline{G}_0. \quad (5)$$

Regression analysis is performed for certain intervals of angles $\overline{\alpha}$. For each interval, the coefficients of the regression model a_1, b_1, c_1, d_1 are calculated using expression [5-7]:

$$B^T = \left(X^T X\right)^{-1}\left(X^T Y\right), \quad (6)$$

where B^T is the column vector of regression coefficients a_1, b_1, c_1, d_1; X is the matrix of independent variables composed of the corresponding values $(\overline{cc})^2 \cdot \sin\overline{\alpha}$, $(\overline{cc})^2$ and $\sin\overline{\alpha}$; Y is the matrix of dependent variables, composed of the corresponding values $\overline{k_T}$.

2.2 Accuracy analysis of the SPS generation forecasting

Accuracy analysis of the forecast is performed by means of determining the forecast error. In this study, WAPE error on hours and WMAPE, % are evaluated [8,9]:

- Weighted absolute percentage error (APE):

$$WAPE = \frac{\left|\overline{G}^m(t) - \overline{G}^f(t)\right|}{\overline{G}_0(t)} \cdot 100\%, \quad (7)$$

where APE is the absolute percentage error, [%]; $\overline{G}^m(t)$ is the average measured value of the solar power flux density near the ground within a defined time step on the time interval t, [W/m^2]; $\overline{G}^f(t)$ is the average forecast value of the solar power flux density near the ground within a defined time step on the time interval t, [W/m^2]; $\overline{G}_0(t)$ is the value of the solar radiation flux density at the boundary of the atmosphere for a defined time step t, [W/m^2].

- Weighted mean absolute percentage error (WMAPE):

$$WMAPE = \frac{1}{N}\sum_{t=1}^{N} \frac{\left|\overline{G}^m(t) - \overline{G}^f(t)\right|}{\overline{G}_0(t)} \cdot 100\%. \quad (8)$$

Estimating the error with regard to the solar radiation flux density at the boundary of the atmosphere will allow estimating the error relating to the conditionally constant "base" and to show how many percent the error is from the maximum possible solar radiation flux density for a defined time interval [10].

3 Calculation example of short-term forecasting of SPS generation

Calculation example presents calculations of short-term forecasts within the period of 04.10.2017 – 21.10.2017 for SPS-1 using different learning samples. In accordance with the proposed clustering methodology, to improve the accuracy of short-term forecasting for the days under consideration, a new sample is formed for each day reflecting the weather conditions.

As part of the data provided by the weather provider, there are: a cloud cover estimate, %; air temperature, ^oC; relative humidity, %; wind speed, m/s. Cloud cover is a key parameter that affects the transparency factor. Thus, it is most expedient to cluster the initial array of retrospective data in accordance with the gradations of cloud cover. The following rules for the formation of blocks of retrospective data were used in the calculations:

- The retrospective includes the values of the corresponding characteristics of x_i and the y response for cloudy days having an average cloud cover within the light day that differs by no more than 30% from the average cloud cover within the light day in the current forecast day.

- The retrospective includes the values of the corresponding characteristics of x_i and the y response for cloudy days having an average cloud cover within the light day differing by not more than 50% of the average cloud cover within the light day in the current forecast day.
- The retrospective includes the values of the corresponding characteristics of x_i and the y response in the range of the minimum and the maximum actual cloud cover for the light day in the current forecast day.
- The retrospective includes the values of the corresponding characteristics of x_i and the y response for all available range.

An example of calculation results is presented for the two most problematic (cloudy) days on October 17-18, 2017 in Table 1.

3.1. Retrospective analysis

In this research, a 6-day retrospective was used to forecast the SPS generation initially for its program implementation. While using a 6-day retrospective in the framework of the presented algorithm, the obtained error is random since the learning sample to calculate the regression coefficients "slides" in 1-day step. In this case, it may occur that the forecast is performed on a cloudy day based on clear days and alternately, which negatively affects the accuracy of short-term forecasting and leads to unreliable results.

In the framework of previous research, the recommendation to use the 6-day retrospective was due to the following factors:

- The amount of initial data on SPS, which was originally provided by the customer for a period of 2 weeks, containing full information about weather conditions and the solar power flux density measurements at SPS.
- The Customer's requirement to minimize the necessary amount of retrospective data to ensure the possibility of rapid implementing the forecasting system at new photovoltaic stations.

3.2. Regression function coefficients

The coefficients of the regression function are chosen in such a way as to minimize the approximation error of the available retrospective data. In the absence of some combination of xi and the y response in the retrospective data, for example, a negative design transparency factor might be obtained when the cloud cover increases to 0.9 – 1.0, which indicates the necessity to expand and/or refine the learning sample.

Therefore, in order to obtain a stable forecast, the learning sample should contain data within the order of 80% for cloudy days, about 20% for clear days. Even when forecasting the solar power flux density for cloudy days, the presence of clear days in retrospective data is necessary to obtain stable coefficients of the regression function.

3.3. Learning sample expanding

The expansion of the learning sample positively affects the accuracy of short-term forecasting. With averaging of the best results obtained within the period of 04.10.2017 – 21.10.2017, the mean absolute percentage error was about 28%. The mean absolute percentage error in the solar power flux density forecast was 49%. The use of the entire available retrospective on average over the period under review provides an error of 38%

For the period under consideration, a number of "problem" days can be identified, when the mean absolute percentage error can reach 100% or more. This problem is due to atypically low transparency factors for the presented cloud conditions. For example, these are 18.10.2017, 17.10.2017.

The retrospective database provided by the Customer does not contain information that uniquely identifies weather conditions; therefore, additional open sources on the Internet have been analyzed to identify cause-effect relationships. At the same time, the database of actual weather conditions of weather providers A and B was considered.

It is remarkable that open databases of actual weather information are not available for specific geographic coordinates and are associated with the locations of weather stations. The closest to the SPS-1 location meteorological station is the Airport-1 weather station, located at a distance of 45 km from SPS-1.

The meteorological station remoteness from the SPS-1 location does not allow using the weather data archive for forecasting, but it provides an opportunity to assess the meteorological situation. Both weather providers A and B provide the same information on the Airport-1 weather station, but B presents more detailed set of meteorological parameters.

For the forecast day of 18.10.2017, broken clouds with a low cloud height of 500 m are characteristic. For the forecast day of 17.10.2017, broken clouds with a low cloud height of 500 m are typical; the type of clouds is cumulonimbus, precipitation is in the form of rain and rain shower (Table 2).

Presented days are characterized by the presence of dense cloud cover and precipitation. As a rule, SPS are built in areas with the greatest number of sunny days per year, therefore the presence of days with precipitation is atypical and the forecast for these days is built with a significant overestimation of the transparency factor, and as a consequence, of the solar power flux density.

Increasing the accuracy of the forecast was achieved by constructing a separate sample of days with precipitation according to the data presented in the archive of Airport-1 on the portal of weather provider B. Thus, the forecast error was reduced: - for 17.10.2017: from 135% to 100%; - for 18.10.2017: from 196% to 75%.

To increase the accuracy of forecasting on days with precipitation, it is necessary to accumulate an appropriate retrospective of the data and to construct a separate sample for days with precipitation. It is necessary to clearly distinguish the type of precipitation: drizzle, rain, rain shower, hail, snow, sleet, etc.

3.4. Revealing the main errors in the learning sample

The greatest contribution to MAPE for the considered day is provided by the hours when the actual data on cloud cover does not correspond to the measured value of the transparency factor. For example, an increase in cloud cover in percent leads to an increase in the transparency factor or alternately. Verification of the data archive issued by the customer was carried out under the following conditions:

- The change in the transparency factor and cloud cover in r.u. are of the same sign; for example, a decrease in cloud cover corresponds to a decrease in the transparency factor.
- Change of one value by more than 0.1 r.u. relative to another does not lead to a change in the second value.

For a sample of 0-20 degrees for the solar altitude angle above the horizon, the number of discrepancies was 137 from the total sample of 503 values (27.2%). For a sample of 20-40 degrees for the solar altitude angle above the horizon, the number of discrepancies was 56 from the total sample of 295 values (18.9%). An estimated evaluation shows that, on average, every fifth change in the actual value of the transparency factor does not correspond to a change in the actual cloud cover. Moreover, in case of taking the forecast cloud cover into account, which is a priori less accurate than the actual one, the error of the initial data increases that introduces a significant error in the calculation results.

The impossibility of identifying weather conditions leads to a high variance of the response y (transparency factor) with respect to the features under consideration x_i (the estimate of cloud cover, the sine of the solar disk altitude angle). For example, Figures 1–3 show the transparency factor function for the angles of the solar disk position above the horizon of 0-20 degrees in the point cloud of the retrospective data (the regression function coefficients were obtained for all the available retrospective data on SPS-1 for angles of 0-20 degrees: $a_1 = 0.3273$, $b_1 = -1.1467$, $c_1 = -0.1404$, $d_1 = 1{,}0551$).

Table 1. Results of the short-term forecast calculations for cloudy days

Date	Characteristic	Retrospective			MAPE, %	WMAPE, %	Retrospective	MAPE, %	WMAPE, %
		Sample size, pcs.		Principle of sampling					
18.10.17	Mean cloud cover (0–20): 0.378 Mean cloud cover (20–40): 0.474	0.0 – 20.0	99	+/- 30% of the actual mean cloud cover per day	94.865	15.518	6 days 0.0 – 20.0: ~96 20.0 – 40.0: ~168	196.095	29.085
		20.0 – 40.0	43						
		0.0 – 20.0	148	+/- 50% of the actual mean cloud cover per day	134.862	19.352			
		20.0 – 40.0	93						
		0.0 – 20.0	194	Based on "Broken clouds" criterion with a cloud height of more than 500 m	75.352	11.138			
		20.0 – 40.0	43						
		0.0 – 20.0	247	in the min/max range of the actual cloud cover within the day	127.012	20.049			
		20.0 – 40.0	203						
		0.0 – 20.0	503	all sample values	135.724	20.777			
		20.0 – 40.0	295						
17.10.17	Mean cloud cover (0–20): 0.940 Mean cloud cover (20–40): 0.977	0.0 – 20.0	265	+/- 30% of the actual mean cloud cover per day	168.7	16.11	6 days 0.0 – 20.0: ~96 20.0 – 40.0: ~168	135.681	13.258
		20.0 – 40.0	138						
		0.0 – 20.0	319	+/- 50% of the actual mean cloud cover per day	155.08	14.724			
		20.0 – 40.0	202						
		0.0 – 20.0	110	Based on "Precipitation", "Fog", "Mist" criteria	101.37	9.314			
		20.0 – 40.0	74						
		0.0 – 20.0	170	in the min/max range of the actual cloud cover within the day	118.309	11.507			
		20.0 – 40.0	75						
		0.0 – 20.0	503	all sample values	138,692	13,352			
		20.0 – 40.0	295						

		20.0 – 40.0	295						

Table 2. The actual weather data of weather provider B

	Date and time	Description of weather events	Cloud characteristic
18.10.2017	18.10.17 14:30	—	Broken (60-90%) 870 m
	18.10.17 14:00	—	Broken (60-90%) 1200 m
	18.10.17 13:30	—	Broken (60-90%) 750 m
	18.10.17 13:00	—	Broken (60-90%) 960 m
	18.10.17 12:30	—	Broken (60-90%) 900 m
	18.10.17 12:00	—	Broken (60-90%) 840 m
	18.10.17 11:30	—	Broken (60-90%) 750 m
	18.10.17 11:00	—	Broken (60-90%) 570 m
	18.10.17 10:30	—	Broken (60-90%) 510 m
	18.10.17 10:00	—	Broken (60-90%) 510 m
	18.10.17 09:30	—	Broken (60-90%) 630 m
	18.10.17 09:00	—	Broken (60-90%) 780 m
	18.10.17 08:30	—	Broken (60-90%) 750 m
	18.10.17 08:00	—	Broken (60-90%) 750 m
	18.10.17 07:30	—	Broken (60-90%) 600 m
17.10.2017	17.10.17 14:30	—	Continuous (100%) 900 m
	17.10.17 14:00	—	Continuous (100%) 780 m
	17.10.17 13:30	—	Continuous (100%) 660 m
	17.10.17 13:00	—	Continuous (100%) 630 m
	17.10.17 12:30	—	Continuous (100%) 630 m
	17.10.17 12:00	—	Continuous (100%) 600 m
	17.10.17 11:30	—	Continuous (100%) 540 m
	17.10.17 11:00	Light rain	Continuous (100%) 480 m
	17.10.17 10:30	Light rain	Continuous (100%) 480 m
	17.10.17 10:00	Light rain	Continuous (100%) 480 m
	17.10.17 09:30	Light rain	Continuous (100%) 480 m
	17.10.17 09:00	Light rain, mist	Continuous (100%) 480 m
	17.10.17 08:30	Light showers, rain	Broken (60-90%) 600 m, cumulonimbus clouds
	17.10.17 08:00	Light showers, rain	Broken (60-90%) 600 m, cumulonimbus clouds
	17.10.17 07:30	Light showers, rain	Broken (60-90%) 600 m, cumulonimbus clouds
	17.10.17 07:00	Light showers, rain	Scattered (40-50%) 480 m, broken (60-90%) 690 m, cumulonimbus clouds

Fig. 1 and 2 show the range of actual transparency factors for a single cloud cover value in percent. Blue, red and green are point clouds in the neighborhood (± 2.5°) of the solar altitude angles above the horizon of 15, 10 and 5 degrees, respectively. The blue, red and green lines depict the functions of the transparency factor at fixed angles of the solar height above the horizon of 15, 10 and 5 degrees, respectively. So, for example, for the Sun angular altitude at 12.5 – 17.5 degrees, the cloud cover value equal to 0.5 corresponds to the transparency factors equal to 0.06, 0.16, 0.25, 0.31, 0.44, 0.50, 0.61, 0.77, etc. The spread is about 70%. The regression function averages the existing retrospective similar to any statistical approach. For the case under consideration, this means impossibility to consider unique weather characteristics for a number of extreme conditions (snow, hail, rain shower, fog, etc.), but increase in the stability of the forecasting system to data exclusions.

It should be noted that the very system of converting the initial data, provided by the weather provider, from the linguistic form to the numeric one by the Customer also helps to reduce the informative content of the learning sample.

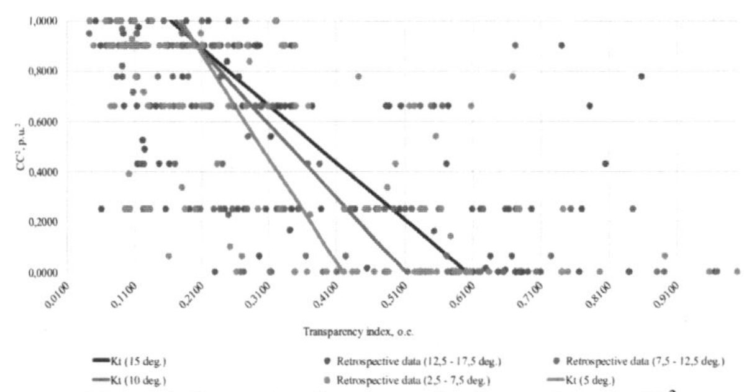

Figure 1. Regression function projection on the axis CC^2

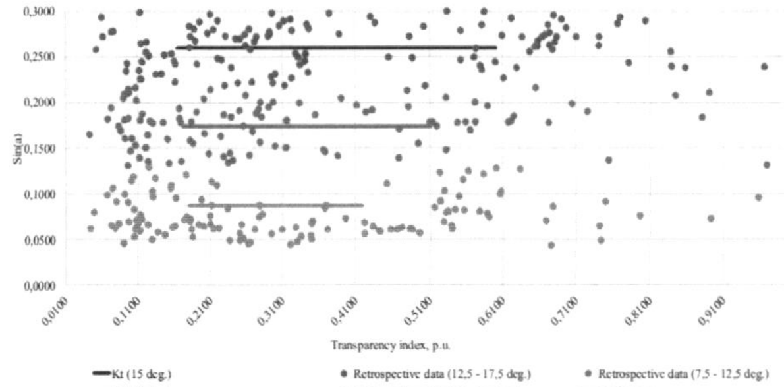

Figure 2. Regression function projection on the axis sinα

4 Conclusion

It is necessary to accumulate additional information on meteorological events and conditions to increase the accuracy of short-term forecasting of SPS generation on cloudy days. It will allow increasing the amount of retrospective and dividing the sampled population into separate samples that will better identify such events as "heavy/light rain", "rain shower", "drizzle", "heavy/light snow", "hail", "fog", "mist", and also classify cloud cover.

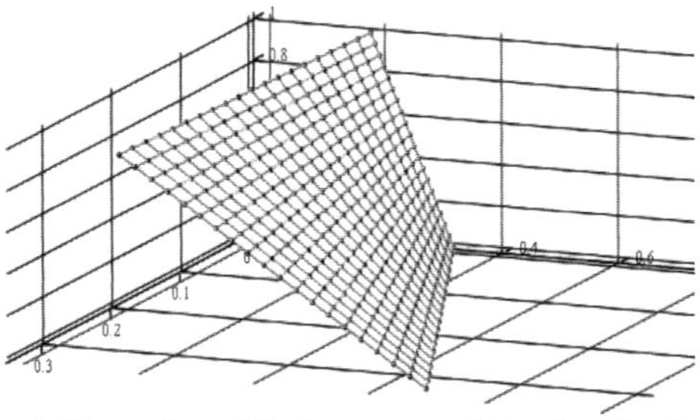

Figure 3. The surface of the transparency factor function for the range of solar altitude angles of 0-20 degrees.

The daily preservation of all possible parameters provided by the weather provider of actual and forecast information (including the original one - the linguistic data provided by the weather provider) is needed for the subsequent improvement of the quality of the SPS generation forecast by reducing the variance of the measured cloud cover values and the calculated transparency factor.

The recommended amount of retrospective allowing accounting the climatic features of the SPS location, as well as clustering of the sampled population into separate samples, is at least 1 year. Provided there are no additional characteristics of weather conditions and events, for short-term forecasting it is recommended:

- To form a retrospective for clear days in the range of an average cloud cover over a day of 0.0 – 0.5 r.u.
- To use the entire available retrospective for cloudy days in the following ratio: 20% of clear days, 80% of cloudy ones.

For new objects (returned to service) it is recommended to use the learning sample according to the rules described above, while ensuring that monthly retrospective record of the data (mostly cloudy days) per sample is made during the current year. The following ratio in the complete sample should be maintained: no more than 20% clear days, not less than 80% of cloudy ones, in addition keeping the total number of pair values (input-output) in the sample of not less than 800 pcs. Record of new retrospective data to the complete sample is made taking into account the recommendations described above, including in its original form - the linguistic data provided by the weather provider.

It is recommended to refine the calculation and display of the forecast error for the short-term SPS generation forecasting and use the APE to estimate the error by hours and the MAPE to estimate the error by day.

References

1. W. Glassley, J. Kleissl. Current state of the art in solar forecasting. California Renewable Energy Collaborative Final Report. 2013.
2. D.A. Snegirev, S.A. Eroshenko, R.T. Valiev, A.I. Khalyasmaa. Algorithmic realization of short-term solar power plant output forecasting. Proceedings of 2017 IEEE 2nd International Conference on Control in Technical Systems, CTS 2017. Pp. 228-231.
3. K.N. Shukla, Saroj Rangnekarb, K. Sudhakar. Comparative study of isotropic and anisotropic sky models to estimate solar radiation incident on tilted surface. Energy Reports 1, pp. 96-103. 2015.
4. W.D. Turner, A. Mujahid. The Estimation of Hourly Global Solar Radiation Using a Cloud Cover Model Developed at Blytheville. American Meteorological Society. 1984, pp. 781-786.
5. M. Paulescu, E. Paulescu, P. Gravila, V. Badescu. Weather Modeling and Forecasting of PV Systems Operation. Springer. 2013.
6. V. Prema, K. Uma Rao. Development of statistical time series models for solar power prediction. Renewable energy. 2015.
7. M. Abuella, B. Chowdhury. Solar Power Probabilistic Forecasting by Using Multiple Linear Regression Analysis. Department of Electrical and Computer Engineering, University of North Carolina, USA. 2015.
8. A. Gondalia, C. Shah. Solar power forecasting analysis of trends in modeling techniques and error minimization mechanism. 2017 2nd IEEE International Conference on Recent Trends in Electronics, Information & Communication Technology (RTEICT), Bangalore, India, 2017, pp. 1860-1866.
9. V. Layanun, S. Suksamosorn, J. Songsiri, "Missing-data imputation for solar irradiance forecasting in Thailand," 2017 56th Annual Conference of the Society of Instrument and Control Engineers of Japan (SICE), Kanazawa, 2017, pp. 1234-1239.
10. M. Alanazi, M. Mahoor, A. Khodaei. Two-stage hybrid day-ahead solar forecasting. 2017 North American Power Symposium (NAPS), Morgantown, WV, 2017, pp. 1-6.

Very-short term solar power generation forecasting based on trend-additive and seasonal-multiplicative smoothing methodology

Stanislav Eroshenko[1], Alexandra Khalyasmaa[1,*] and Rustam Valiev[1]

[1]Ural Federal University named after the first President of Russia B.N. Yeltsin, Ekaterinburg, 620002 Mira str. 19, Russia

Abstract. In conditions of development of generating facilities on renewable energy sources, the technology runs up to uncertainty in the operational and short-term planning of the power system operating modes. To date, reliable tools for forecasting the generation of solar power stations are required. This paper considers the methodology of operational forecasting of solar power stations output based on the mathematical apparatus of cubic exponential smoothing with trend and seasonal components. The presented methodology was tested based on the measuring data of a real solar power station. The average forecast error was not more than 10% for days with variable clouds and not more than 3% for clear days, which indicates the effectiveness of the proposed approach.

1 Introduction

In conditions of the energy market development, the role of effective management tasks at all levels – the power station, the power system, the power pool system – is immeasurably increasing. At the same time, the effectiveness of the operational optimal control of the power system mode is firstly determined by the correctness and accuracy of the forecast for the power stations generation and the load power. The most important procedures in the power joint market are the competitive capacity auction on the day-ahead market, solving the technical and economic problem to select the live generating equipment, as well as the calculation and considering generation patterns under the power system actual control on-line within the balancing power market.

In accordance with the rules for the price bids submission by participants in the wholesale market [1], the applications of electricity suppliers are submitted taking into account following requirements:

- applications on the day-ahead market are sent for every hour of the day *X* (applications are sent per day *X-1* to 13:30 Moscow time);
- applications on the balancing market are sent every hour (one hour before calculating the balancing market plan).

In conditions of development of generating facilities on renewable energy sources, the degree of uncertainty in the operational and short-term planning of the power system operating modes is significantly increased. To date, reliable tools for forecasting the generation of power stations using, in particular, solar energy are required. Taking into account the nature of the wholesale electricity market in the UES of Russia, the short-term forecasting of solar power stations generation is considered in terms of the application on the day-ahead market. It is also required to obtain reference data on the solar power station generation for the day *X+1, X+2,..., X+N*, depending on the availability of relevant weather forecast data provided by the weather provider.

When controlling the operating mode of the power system on-line, it is necessary to compensate the departure from the mode planned one day ago. Therefore, the task of operational forecasting is considered in terms of filing applications on a balancing market for the next hour, as well as for obtaining reference data on the solar power station generation for several hours ahead.

To date, there are many methods for forecasting the generation of solar power stations based on the forecasting the solar radiation flux density. There are a number of outstanding methodologies based on the identification of satellites images of cloud movement in the area where the object under consideration is located [2].

In a number of works, it is proposed to use specialized measuring instruments that allow quickly assessing the weather conditions. The use of these methods, as a rule, is limited by the characteristics of the weather data being analyzed, or requires additional investments for the installation of meteorological measuring posts. From the point of view of implementation, statistical methods of time series analysis, such as moving average method, autoregressive models, etc. are the most simple [3,4]. In this paper, the application of exponential smoothing methods to solve the problem of solar power stations operational forecasting is considered.

For small intervals from 1 to 3 hours, the generation forecast can be significantly improved by using current

*Corresponding author: lkhalyasmaa@mail.ru

data obtained by direct observation methods. The main requirement for measuring current weather conditions at a solar power station is the horizontally installed pyranometer for measuring the total flux density of solar radiation. Current measurements obtained from the pyranometer are used in the algorithm of operational forecasting of solar power stations.

The accuracy of the forecast, as a rule, is estimated by the deviations between the forecast and actual values; the following attributes are often used to estimate accuracy [5]:

- mean absolute error value – is useful in assessing the possible penalties associated with the forecast inaccuracy;
- mean-square error value – more focuses on large departures of the forecast from actual values;
- systematic error values associated with underestimation or revaluation of output.

2 Methodology description

The operational forecast is implemented to control the operation of the solar power station (5-120 minutes) and is actually a refinement of the short-term forecast adjusted for current measurement data. Forecasting for relatively small prediction intervals is effectively implemented based on statistical methods of time series analysis, in particular, the exponential smoothing method [6].

Exponential smoothing is one of the most common methods of leveling the series. It is based on the calculation of exponential averages. Exponential smoothing of the series is carried out according to the recurrence formula:

$$S_t = \alpha \cdot y_{t-1} + (1-\alpha) \cdot S_{t-1}, \qquad (1)$$

where S_t is the exponential average at the time t; α is the smoothing parameter ($\alpha = \text{const}, 0 < \alpha < 1$); y_{t-1} is the actual value of the solar radiation flux density at the time $t-1$; S_{t-1} is the exponential average at the time $t-1$.

If the recurrence relation (1) is used successively, the exponential average S_t can be expressed in terms of the values of the series y :

$$S_t = \alpha \sum_{i=0}^{N-1} (1-\alpha)^i y_{t-i} + \beta_N S_0, \qquad (2)$$

where S_t is the exponential average at the time t ; N is the number of series terms; α is the smoothing parameter ($\alpha = \text{const}, 0 < \alpha < 1$); y_{t-i} is the actual value of the solar radiation flux density at the time $t-i$; β_N is the parameter for smoothing the initial conditions; S_0 is the value of the solar radiation flux density, which characterizes the initial conditions for the first application of formula (1) for $t = 0$.

As the initial approximation, the value S_t turns out to be the weighted sum of all the terms of a series consisting of average values of the solar radiation flux density, and the weights decrease exponentially as a function of the prescription ("age") of the observation.

Exponential alignment always requires a previous exponential average. When the process is beginning, there must be some value S_0 that can be used as the value preceding S_1 . Thus, if there are past data on the flux density of solar radiation by the time of the beginning of the alignment process, then the arithmetic mean of all available points or some part of them can be used as an initial value S_0.

Identification of the initial level of the series S_0 can also be performed based on a priori knowledge about the process of changing the flux density of solar radiation during the day or based on its analogy with other processes. After k steps, the weight given to the initial value is $(1-\alpha)^k$. If there is confidence in the validity of the initial value S_0, then the coefficient α can be taken as small. If there is no such confidence, the parameter α should be given a large value, so that the influence of the initial value decreases rapidly. However, a large value of α can cause a large dispersion of S_t oscillation. If suppression of these oscillations is required, then after sufficient removal from the initial time moment, the value α can be reduced.

The described approach is an exponential smoothing of the first order, which is mainly applicable for simple data without pronounced trends or seasonal changes, but when the data includes these components, it is recommended to use exponential smoothing of the second and third degree for more accurate forecast. Since the value predicted in the work described - the flux density of solar radiation - has seasonals, it is recommended to use cubic exponential smoothing. Basic equations for its implementation are presented below:

$$P_{f(t)} = \alpha \cdot \frac{P_{t-1}}{I_{t-L}} + (1-\alpha) \cdot (P_{f(t-1)} + b_{t-1}), \qquad (3)$$

where $P_{f(t)}$ is the exponential average at the time t ; α is the smoothing parameter ($\alpha = \text{const}, 0 < \alpha < 1$); P_{t-1} is the actual value of the solar radiation flux density at the time $t-1$; b_{t-1} is the trend component at time $t-1$; I_{t-L} is the indicator that takes into account seasonals in the flux density of solar radiation.

The trend component for the time t is determined by the formula:

$$b_t = \gamma(P_{f(t)} - P_{f(t-1)}) + (1-\gamma)b_{t-1}, \quad (4)$$

where b_t is the trend component at time t; γ is the smoothing parameter ($\gamma = \text{const}, 0 < \gamma < 1$); $P_{f(t)}$ is the exponential average at the time t; $P_{f(t-1)}$ is the exponential average at the time $t-1$; is the trend component at time $t-1$.

The indicator taking into account seasonals of the predicted value is determined by the formula:

$$I_t = \beta \cdot \frac{P_t}{P_{f(t)}} + (1-\beta)\cdot I_{t-L} \cdot \gamma, \quad (5)$$

where I_t is the indicator taking into account seasonals of the predicted value for the current season; β is the smoothing parameter ($\beta = \text{const}, 0 < \beta < 1$); P_t is the actual value of the predicted value at the time t; $P_{f(t)}$ is the exponential average at the time t; I_{t-L} is the indicator taking into account seasonals of the predicted value for the previous season; γ is the smoothing parameter ($\gamma = \text{const}, 0 < \gamma < 1$).

It is possible to determine the predicted value for m intervals ahead based on the trend component for the time t and the indicator that takes into account seasonals of the predicted value:

$$P_{f(t)t+m} = (P_{f(t)} + m \cdot b_t) \cdot I_{t-L+m}, \quad (6)$$

where $P_{f(t)t+m}$ is the predicted value for m intervals ahead; $P_{f(t)}$ is the exponential average at the time t; b_t is the trend component at time t; I_{t-L+m} is the indicator taking into account seasonals of the predicted value within the season for m intervals ahead.

To use the model, initialization of the parameters at the initial stage is necessary. If the forecasted data includes N periods, then in order to estimate the trend component it is recommended to use the initial retrospective data of at least $2N$ periods. The equation for the initial estimate of the indicator b is as following:

$$b = \frac{1}{N}\left(\frac{P_{N+1} - P_1}{N} + \frac{P_{N+2} - P_2}{N} + \ldots + \frac{P_{N+N} - P_N}{N}\right) \quad (7)$$

It is also necessary to calculate the initial characteristics of the seasonals effect. Assume that there are data on the solar radiation flux density for 6 years, with data for 4 seasons for each year. At the first stage, the average value of the predicted parameter is calculated for each year y:

$$A_y = \frac{\sum_{i=1}^{4} P_i}{4}, \; y = 1, 2, 3 \cdots 6 \quad (8)$$

At the second stage, the specific weights of the influence of each season for each year are estimated by dividing the actual values of the predicted parameter by the average for each year.

$$\begin{array}{cccccc} 1 & 2 & 3 & 4 & 5 & 6 \\ P_1/A_1 & P_5/A_2 & P_9/A_3 & P_{13}/A_4 & P_{17}/A_5 & P_{21}/A_6 \\ P_2/A_1 & P_6/A_2 & P_{10}/A_3 & P_{14}/A_4 & P_{18}/A_5 & P_{22}/A_6 \\ P_3/A_1 & P_7/A_2 & P_{11}/A_3 & P_{15}/A_4 & P_{19}/A_5 & P_{23}/A_6 \\ P_4/A_1 & P_8/A_2 & P_{12}/A_3 & P_{16}/A_4 & P_{20}/A_5 & P_{24}/A_6 \end{array} \quad (9)$$

The obtained results are used in the third stage to find seasonal coefficients as the arithmetic mean of the calculated values.

$$\begin{aligned} I_1 &= \frac{P_1/A_1 + P_5/A_2 + P_9/A_3 + P_{13}/A_4 + P_{17}/A_5 + P_{21}/A_6}{6} \\ I_2 &= \frac{P_2/A_1 + P_6/A_2 + P_{10}/A_3 + P_{14}/A_4 + P_{18}/A_5 + P_{22}/A_6}{6} \\ I_3 &= \frac{P_3/A_1 + P_7/A_2 + P_{11}/A_3 + P_{15}/A_4 + P_{19}/A_5 + P_{23}/A_6}{6} \\ I_4 &= \frac{P_4/A_1 + P_8/A_2 + P_{12}/A_3 + P_{16}/A_4 + P_{20}/A_5 + P_{24}/A_6}{6} \end{aligned} \quad (10)$$

The special attention should be paid to choosing the value of smoothing constants α, β and γ. As a rule, these coefficients tend to be fit in such a way as to minimize the mean square deviation of the forecast from the actual measurements. The use of the presented model requires the availability of retrospective data, while reliable estimation of seasonals and trend components requires information on the predicted parameters for at least two seasons.

The exponential smoothing method Implementation for operational forecasting of solar power stations generation is possible with the following:

1. The development of the cubic exponential smoothing model taking into account the trend component and seasonals of the solar radiation flux density in the first stages of implementation can be limited by a model of the first or second order by the moment of accumulation of retrospective data on solar radiation flux density.

2. In order to ensure a unified approach to the forecasting problem, the function of calculating the operational forecast is expedient to be implemented for estimating the forecast of the solar radiation flux density. In this case, the solar power station generation can be found due to a known number and passport data of photovoltaic panels, inverters, converters and primary equipment on alternating current.

3. Taking into account the stochastic nature of the considered process and the presence of large "drops" in the time series of the solar flux density during periods of variable clouds in order to ensure the exponential smoothing method adaptability to the current design conditions, smoothing coefficients α, β and γ will be found automatically for each current design day X based on the previously calculated short-term forecast of the solar radiation flux density.

Input data: short-term forecast of the solar radiation flux density for the current day X taking into account the updated weather forecast provided by the weather provider $\left\{\widehat{G}_{h0}, \widehat{G}_{h1}, \ldots \widehat{G}_{h24}\right\}$.

Output data: an operational correction of the forecast of the solar flux density for the current day X taking into account the weather data received from the meteorological station established on the territory of the solar power station $\left\{G_{h0}, G_{h1}, \ldots, G_{ht}, \widehat{G}_{h\,t+1} \ldots, \widehat{G}_{h24}\right\}$.

Fitting the coefficients of the exponential smoothing function α, β and γ is carried out till the actual sunrise time of the current day, based on the data of the short-term forecast of the solar flux density for the current day by estimating the weighted mean percentage error:

$$WMPE = \frac{1}{N}\sum_{t=1}^{N} \frac{\left|\overline{G}^{m}(t) - \overline{G}^{f}(t)\right|}{\overline{G}_0(t)} \cdot 100\%, \qquad (11)$$

where $\overline{G}^{m}(t)$ is the average measured value of the solar radiation flux density near the ground within the defined time step on the time interval t, [W/m^2]; $\overline{G}^{f}(t)$ is the average forecast value of the solar radiation flux density near the ground within the defined time step on the time interval t, [W/m^2]; $\overline{G}_0(t)$ is the solar radiation flux density at the boundary of the atmosphere for the defined time step t, [W/m^2]. The flux density of solar radiation at the boundary of the atmosphere is determined in accordance with the expression [7]:

$$\overline{G}_0 = \frac{12}{\pi} G_{on} (\cos\varphi \cdot \cos\delta \cdot (\sin\omega_2 - \sin\omega_1) + \frac{\pi(\omega_2 - \omega_1)}{180^0} \cdot \sin\varphi \cdot \sin\delta), \qquad (12)$$

where $\overline{G}_0$ is the average value of the solar radiation flux density at the boundary of the atmosphere within the defined time step, [W/m^2]; G_{on} is the solar radiation flux density at the boundary of the atmosphere, [W/m^2]; φ is the latitude, [degrees]; δ is the solar declination angle, [degrees]; ω_1 is the hour angle at the beginning of the interval, [degrees]; ω_2 is the hour angle at the end of the interval, [degrees].

The weighting of the error at the boundary of the atmosphere will allow estimating the error relative to the conditionally constant "base" and showing how many percent the error is from the maximum possible flux density of solar radiation for the considered time interval [8].

This criterion is resistant to individual "spikes" of the model and provides the ability to fit automatically optimal smoothing coefficients for the current design day X by the brute force, which is easily implemented programmatically.

3 Calculation example

Within the framework of the presented work, the feasibility of implementing the operational forecast function for the existing solar power station was analyzed. Geographic parameters, as well as the time zone parameters are presented in the Table 1.

Table 1. Geographic parameters of the solar power station location

№	Parameter	Symbol	Units	Value
1.	Latitude	b_1	[degrees]	51.26
2.	Longitude	c_1	[degrees]	58.47
3.	Difference with Greenwich time	d_1	[hours]	5

The data of current measurements for the period from 01.05.2017 to 28.05.2017 were taken for the analysis. Measuring data on the flux density of solar radiation is represented by averaged half-hour values. Since within the balancing market the information is submitted 1 hour before the planning of the power system, the forecasting of the solar power stations generation is 1 and 2 steps ahead. Their averaged value is taken as their predicted value.

The forecast error parameters for the characteristic days of the period under consideration are presented in Table 2.

Table 2. The forecast error in case of cubic exponential smoothing

№	Date	Weather conditions	Weighted error for *30 minutes ahead, %*	Weighted error for *an hour ahead, %*
1.	01.05.17	clear (no clouds)	2.77	4.03
2.	03.05.17	light clouds	4.36	5.52
3.	19.05.17	variable clouds	7.35	9.49
4.	11.05.17	continuous clouds	5.98	7.16
5.	01.05.2017 – 28.05.2017		4.96 min – 1.87 max – 9.85	6.50 min – 2.45 max – 12.69

Graphical representation of the actual and forecast curve for some characteristic days of the period under consideration is presented in Figures 1-2.

To illustrate the effectiveness of the methods and approaches for operational forecasting of solar power generation, Figure 3 presents a graphical comparison of the accuracy of the short-term solar power station generation forecast for the day ahead and operational forecast.

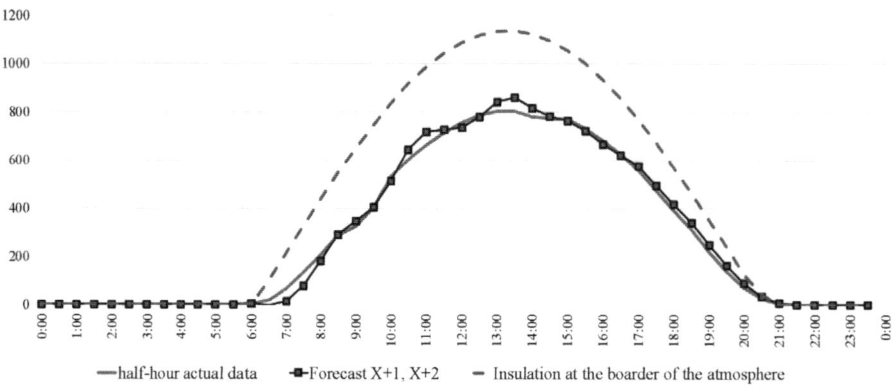

Figure 1. Hour-ahead exponential smoothing forecasting, 01.05.2017

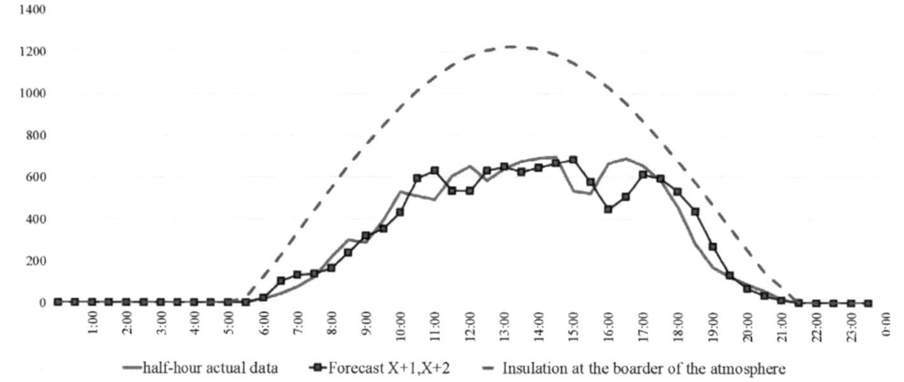

Figure 2. Hour-ahead exponential smoothing forecasting, 19.05.2017

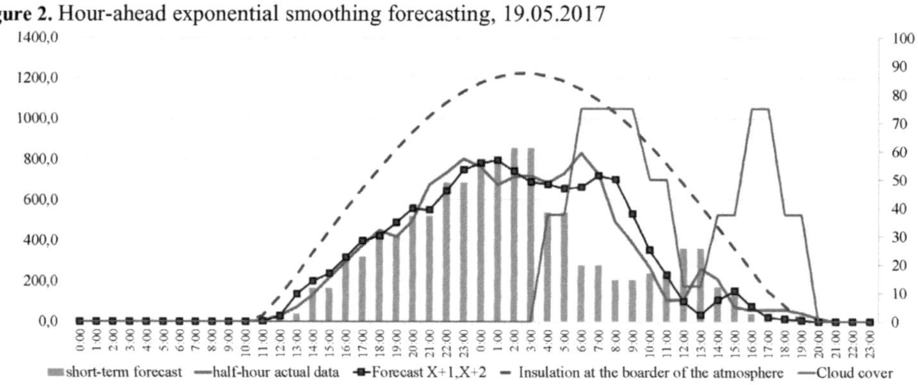

Figure 3. Hour-ahead exponential smoothing forecasting, 28.05.2017

Obviously, an operational correction based on current measurements allows to significantly reduce the error, which is due, among other things, to the low accuracy of weather data, on the basis of which the day ahead forecast was made.

4 Conclusion

In this paper an approach to operational forecasting of solar power generation based on the cubic exponential smoothing method, taking into account the trend and seasonal components in the initial sample, was considered.

Accounting the seasonal component, obtained by the methodology presented in this article, allows to significantly reduce the "backlog" effect typical for all forecasting methods based on the retrospective data analysis, thereby accuracy of forecasting on clear and cloudy days significantly increases.

The average error calculated with respect to the flux density of solar radiation at the boundary of the atmosphere was no more than 10%, which proves the high accuracy of the forecast of output of the generating object. It should be noted that the error substantially increases as the forecast horizon deepens, which indicates the necessity to revise this methodology when predicting 3 or more hours ahead.

Greater accuracy can be achieved by increasing the discreteness of the information analyzed. Therefore, when predicting with a discreteness of 30 minutes, it is expedient to monitor the dynamics of the predicted parameter at 10-15 minutes intervals, which allows to correct the trend component promptly.

References

1. V. Margaret, J. Jose. Exponential Smoothing Models for Prediction of Solar Irradiance. International Journal of Advanced Research in Electrical, Electronics and Instrumentation Engineering (An ISO 3297: 2007 Certified Organization) Vol. 4, Issue 2, February 2015.
2. D.A. Snegirev, S.A. Eroshenko, R.T. Valiev, A.I. Khalyasmaa. Algorithmic realization of short-term solar power plant output forecasting. Proceedings of 2017 IEEE 2nd International Conference on Control in Technical Systems, CTS 2017. Pp. 228-231.
3. C. Severiano, F. G. Guimarães and M. W. Cohen, "Very short-term solar forecasting using multi-agent system based on Extreme Learning Machines and data clustering," 2016 IEEE Symposium Series on Computational Intelligence (SSCI), Athens, 2016, pp. 1-8.
4. C.A. Severiano, P. C. L. Silva, H. J. Sadaei and F. G. Guimarães, "Very short-term solar forecasting using fuzzy time series," 2017 IEEE International Conference on Fuzzy Systems (FUZZ-IEEE), Naples, 2017, pp. 1-6.
5. A.D. Orjuela-Cañón, J. Hernández and C. R. Rivero, "Very short term forecasting in global solar irradiance using linear and nonlinear models," 2017 IEEE Workshop on Power Electronics and Power Quality Applications (PEPQA), Bogota, 2017, pp. 1-5.
6. S. Gupta, N. A. Shrivastava, A. Khosravi and B. K. Panigrahi, "Wind ramp event prediction with parallelized gradient boosted regression trees," 2016 International Joint Conference on Neural Networks (IJCNN), Vancouver, BC, 2016, pp. 5296-5301.
7. W. Glassley, J. Kleissl. Current state of the art in solar forecasting. California Renewable Energy Collaborative Final Report. 2013.
8. V. Prema. K. Uma Rao. Development of statistical time series models for solar power prediction. Renewable energy. 2015.

Machine learning techniques for short-term solar power stations operational mode planning

Stanislav Eroshenko[1], Alexandra Khalyasmaa[1,*] and Denis Snegirev[1]

[1]Ural Federal University named after the first President of Russia B.N. Yeltsin, Ekaterinburg, 620002 Mira str. 19, Russia

Abstract. The paper presents the operational model of very-short term solar power stations (SPS) generation forecasting developed by the authors, based on weather information and built into the existing software product as a separate module for SPS operational forecasting. It was revealed that one of the optimal mathematical methods for SPS generation operational forecasting is gradient boosting on decision trees. The paper describes the basic principles of operational forecasting based on the boosting of decision trees, the main advantages and disadvantages of implementing this algorithm. Moreover, this paper presents an example of this algorithm implementation being analyzed using the example of data analysis and forecasting the generation of the existing SPS.

1 Introduction

At present, machine-learning algorithms that use decision trees are very common and universal for most applications. They allow predicting the real response for each object, that is, solving the regression problem. Composing several derived trees, combining the responses of each of them, it is possible to get a stable and much more qualitative solution than many other algorithms can provide.

Forecasting the generation of a solar power station (SPS), both short-term (day ahead) and operational (one hour ahead), consists of five main stages:

- Determining the solar radiation flux density (SRFD) at the boundary of the atmosphere;
- Determining the SRFD incident on the horizontal surface of the earth;
- Determining the SRFD incident on the inclined panel surface;
- Determining the generation of a photoelectric DC converter;
- Determining the power output of AC inverters.

The most difficult part of the SPS generation forecasting process is the second stage, which is the determination of SRFD incident on the horizontal surface of the earth, since this value depends on a variety of unstable and difficult for forecasting factors, the greatest influence of which is exerted by cloud cover.

In case of short-term forecasting, the dependence of SRFD incident on the horizontal surface of the earth on cloud cover can be restored by solving the regression problem based on least-squares method (or other similar method) and choosing the optimal function [1,2]. The use of such traditional methods to solve the regression problem for operational forecasting of SRFD incident on the horizontal surface of the earth is not possible, since in the case of variable clouds it is not possible to restore the relationship between SRFD and the current measurement data, i.e. variables that allow describing this dependence.

2 SPS operational forecasting model

2.1 Problem formulation

In this research the task of SRFD operational forecasting using retrospective data was set, based on which the output power of the SPS was calculated [3]. There is a learning sample S :

$$S = \begin{bmatrix} x_{11} & x_{21} & \cdots & x_{1a} & y_1 \\ x_{12} & x_{22} & \cdots & x_{2a} & y_2 \\ \cdots & \cdots & \cdots & \cdots & \cdots \\ x_{l1} & x_{l2} & \cdots & x_{la} & y_l \end{bmatrix}, \quad (1)$$

where y_j is the parameter, which is the measured value of SPFD within the task; x_{ij} is the attribute matching the parameter y_i (within the task, attributes can be both calculated values, solar declination angle δ , SRFD on the boundary of atmosphere $\overline{G}_0$, and retrospective parameters, such as transparency factor $\overline{k}_T$ as well as weather data); l is the number of observations in the sample; a is the number of attributes.

The task consisted in composing the decision trees, which with great accuracy will determine the value of the new parameters y according to the relevant attributes x_{ij} [4,5]; in other words, the task consisted in development of a *model* (function) f that, having received x on input,

*Corresponding author: lkhalyasmaa@mail.ru

would predict the value of the response y . In this research, a gradient-boosting algorithm was used with decision trees.

2.2 Gradient boosting

Gradient boosting is essentially a gradient descent in the space of all possible algorithms [6]; each step of this descent is done by the basic algorithm according to the following sequence of actions $b_n(x)$ according to the following sequence of actions.

- Initializing of the first basic algorithm b_0

$$b_0(x) = 0. \quad (2)$$

- For $n = 1,\ldots,N$ the following steps are repeated:
  1. The shift vector s , which shows how to correct the predictions of the composition already constructed to reduce the error in the learning sample, is calculated:

$$s_n = \left(-2(a_{n-1}(x_1) - y_1),\ldots,-2(a_{n-1}(x_l) - y_l)\right) \quad (3)$$

  2. A basic algorithm b_n is constructed by approximating its responses to the learning sample to the shift data s_i :

$$b_N(x) = \arg\min_b \frac{1}{l}\sum_{i=1}^{l}\left(b(x_i) - s_i\right)^2 = \sum_{j=1}^{J}\left[x \in R_{Nj}\right]b_{Nj} \quad (4)$$

  3. After the algorithm is found, it is added to the composition:

$$a_n(x) = a_{n-1}(x) + \eta\sum_{j=1}^{J}\left[x \in R_{Nj}\right]b_{Nj} \quad (5)$$

  4. Steps 1-3 are repeated until a stop criterion, for example, a specified number of iterations, is satisfied.

The settings shown in Table 1 were used to work with the gradient-boosting algorithm. The *colsample_bytree* and *colsample_bylevel* settings were used to avoid overfit. To solve the problem, authors developed an approach based on gradient boosting on decision trees. The task was proposed to be solved in three ways: teaching an algorithm without history, teaching only with history and teaching with history and weather data.

Computational experiments were performed to evaluate the possibility of using the gradient-boosting algorithm on solution trees for the task of SPS generation operational forecasting within the research work. Experiments varied according to the following criteria:

- Teaching without history

The following calculated values were used as attributes:

1. Number of a day in a year n ;
2. Solar declination angle δ , degrees;
3. Solar time t_s , hours;
4. SRFD at the boundary of the atmosphere G_0 , W/m^2.

- Teaching with history

As attributes, the calculated values and the data obtained by measurements were used:

1. Transparency factor $k_T = \frac{\overline{G}_m}{\overline{G}_0}$, p.u.;
2. Number of a day in a year n ;
3. Solar declination angle δ , degrees;
4. Solar time t_s , hours;
5. SRFD at the boundary of the atmosphere G_0 , W/m2.

- Teaching with history and weather data

As attributes, the calculated values, the data obtained by measurements, as well as actual weather data were used:

1. Transparency factor $k_T = \frac{\overline{G}_m}{\overline{G}_0}$, p.u.;
2. Average temperature per hour temp , K;
3. Maximum temperature per hour temp_max , K;
4. Minimum temperature per hour temp_min , K;
5. Average pressure per hour pressure , hPa;
6. Average humidity per hour humidity , %;
7. Average wind speed per hour wind_speed , m/s;
8. Average wind direction per hour wind_deg , degrees;
9. Precipitation in the last 3 hours rain_3h , mm;
10. Average cloud cover per hour clouds_all , %;
11. Weather identifier (presence or absence of cloud cover), weather_id ;
12. Number of a day in a year n ;
13. Solar declination angle δ , degrees;
14. Solar time t_s , hours;
15. SRFD at the boundary of the atmosphere G_0 , W/m2.

Table 1. Gradient boosting configuration

Configuration options	Types of input data		
	Teaching without history	Teaching with history	Teaching with history and weather data
Step length (*learning rate*) η	0.005	0.01	0.01
Trees maximum depth (*max_depth*)	4	4	6
Number of trees (*n_estimators*)	3000	3000	3000
Number of iterations (*n_boost_round*) N	3000	3000	3000
Share of variables used at each iteration (*colsample_bytree*)	0.8	0.8	0.8
Share of variables used at each level of the tree (*colsample_bylevel*)	0.7	0.7	0.7
Threshold value G_{thres}	10	10	10

2.3 Developed model evaluation

A cross-validation was used to evaluate the analytical model and its behavior on independent data. The available data were divided into 5 parts for the model evaluation. The algorithm for the learning subsample is configured for each partition; then its mean error at the objects of the control subsample was estimated. The cross-validation estimation was the error average for all partitions on the control subsamples [6].

To analyze the prediction error, the mean absolute percentage error MAPE and the mean-square error (MSE) were used [7,8]. The mean absolute percentage error (MAPE) was calculated by the formula:

$$MAPE = \frac{1}{l}\sum_{i=1}^{l}\frac{\left|\overline{G}_m^i - \overline{G}_f^i\right|}{\overline{G}_m^i}\cdot 100\%, \qquad (6)$$

where $MAPE$ is the mean absolute percentage error, [%]; l is the number of objects in the sample; $\overline{G}^m$ is the average measured value of the SRFD near the ground, [W/m^2]; $\overline{G}^f$ is the average forecast value of the SRFD near the ground, [W/m^2]. $MAPE$ does not allow adequately estimating the error in the range of small values $\overline{G}_m$ so a threshold value $\overline{G}_m > 100$ is used to calculate $MAPE$.

Mean-square error (MSE) is calculated by the formula:

$$MSE = \frac{1}{l}\sum_{i=1}^{N}\left(\overline{G}_m^i - \overline{G}_f^i\right)^2, \qquad (7)$$

where MSE is the mean-square error, [W^2/m^4]; N is the number of time intervals t in the day considered (between sunrise and sunset); $\overline{G}^m$ is the average measured value of the SRFD near the ground, [W/m^2]; $\overline{G}^f$ is the average forecast value of the SRFD near the ground, [W/m^2].

3 Calculation example of SPS operational forecasting

The results of the operational forecasting correction for the SRFD incident on the horizontal surface of the earth $\overline{G}$ for all three teaching modes in case of the 1-hour forecast horizon are shown in Fig. 1 - 2.

Parameters of the forecast error for considered days are presented in the Table 2.

It is seen from the Figures 1 and 2 that in case of teaching without history (without using current measurement data) the algorithm allows determining the relationship between SRFD at the boundary of the atmosphere $\overline{G}_0$ and SRFD incident on the horizontal surface of the earth $\overline{G}$ without considering cloud cover. This teaching mode physically takes into account only the change in the solar altitude angle during the day, as well as factors that are generally constant (solar absorption by ozone, dust, air molecules, water).

In case of teaching with history, when the nearest available data of current measurements were used for the prediction, the algorithm allows to approximate the forecast value $\overline{G}_f$ to the measured values $\overline{G}_m$. For the case of slightly variable clouds on 28.05.2017, the algorithm allows to correct the forecast even without reliance on weather data. For the case of sharply variable clouds on 27.05.2017, an accurate correction was not possible; the error for 8 of 16 hours was 50% or more.

The operational forecast correction for such days necessitates the use of reliable weather data. A decrease in the prediction error is observed for both days considered in case of teaching with history and weather data, even though low quality weather data were used to teach the algorithm.

Table 3 shows the parameters of the forecast error for the entire sample considered. This data is from 14.01.2017 to 28.05.2017. It can be seen from the table that the proposed algorithm has a lower average error for the 1-hour forecast horizon within the entire considered time interval (4.5 months) than for a single day. This fact points at the significant accuracy of operational forecasting and the success of the proposed algorithm application for most of the considered days.

From Table 3 it is also clear that the use of low quality weather data provided only a slight error reduction compared with teaching only with history. Obviously, using more accurate weather data will result in making more accurate operational forecasts.

The main factor affecting the error of operational forecast is the remoteness of the forecast horizon. Figures 3 and 4 show the dependence of the mean-square error MSE and the mean absolute percentage error $MAPE$ on the magnitude of the forecast horizon. It can be seen from the diagrams that the accuracy increases as the forecast horizon approaches, since the current measurements are more informative (better describe the predicted value).

From the analysis of the results obtained, it can be concluded that it is possible to apply the gradient-boosting algorithm on solution trees in the context of the task of SPS generation operational forecasting.

Table 2. Forecast error parameters ($H = 1\ hour$)

Day	Type of input data					
	Without history		With history		With history and weather data	
	MAPE, %	*MSE*, (W/m^2)^2	*MAPE*, %	*MSE*, (W/m^2)^2	*MAPE*, %	*MSE*, (W/m^2)^2
27.05.17	91,01	47514,50	59,03	24835,58	49,89	24461,12
28.05.17	26,56	10056,30	22,96	9224,44	18,46	9720,42

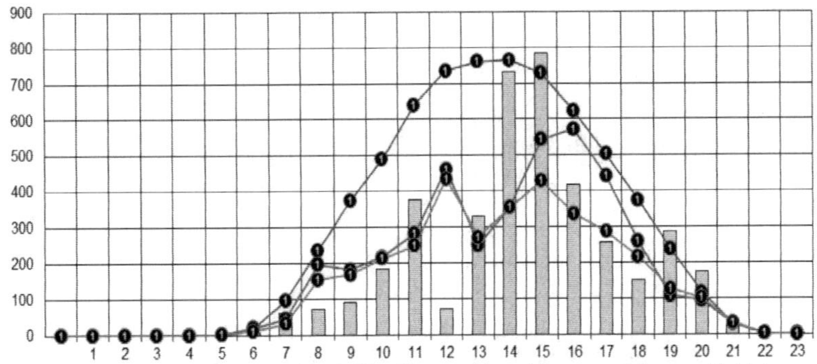

Figure 1. 1-hour ahead operational forecast for 27.05.2017 (blue line – without history, red line – with history, orange line – history and weather)

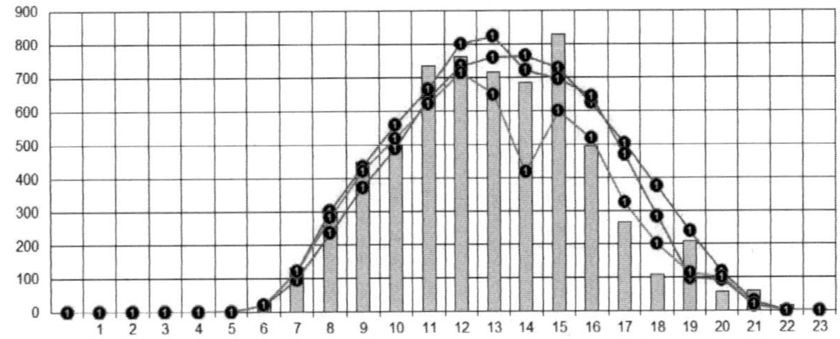

Figure 2. 1-hour ahead operational forecast for 28.05.2017 (blue line – without history, red line – with history, orange line – history and weather)

Table 3. Forecast error parameters for the entire sample considered

Forecast horizon H	Types of input data					
	Without history		With history		With history and weather data	
	MAPE, %	*MSE*, $(W/m^2)^2$	*MAPE*, %	*MSE*, $(W/m^2)^2$	*MAPE*, %	*MSE*, $(W/m^2)^2$
1	45.84	15950.74	21.06	3651.02	19.88	3363.92
2	45.84	15950.74	28.23	6092.28	24.94	4904.88
3	45.84	15950.74	32.77	8188.28	28.09	5908.13
4	45.84	15950.74	36.77	10476.08	30.63	6627.15
5	45.84	15950.74	40.25	12644.29	32.46	7474.96
6	45.84	15950.74	43.11	14272.88	35.01	8896.09
7	45.84	15950.74	44.82	15489.63	38.07	10589.30
8	45.84	15950.74	45.80	16235.96	39.60	11424.11
9	45.84	15950.74	46.48	16630.79	40.82	11943.67
10	45.84	15950.74	46.34	16556.05	41.48	12665.08
11	45.84	15950.74	46.78	16852.75	41.19	12596.49
12	45.84	15950.74	46.46	16733.06	42.63	13686.00
13	45.84	15950.74	46.64	17000.71	43.40	14260.52
14	45.84	15950.74	46.36	16646.83	44.39	15398.75
15	45.84	15950.74	45.73	16042.63	44.15	15073.16
16	45.84	15950.74	45.61	15718.92	44.47	15030.38
17	45.84	15950.74	45.22	15446.75	45.45	15741.26
18	45.84	15950.74	45.75	15794.79	45.19	15567.77
19	45.84	15950.74	45.52	15842.28	46.40	16543.24
20	45.84	15950.74	45.41	15923.70	47.10	17016.06
21	45.84	15950.74	45.50	16186.40	46.43	16885.56
22	45.84	15950.74	45.81	16655.97	46.38	17081.73
23	45.84	15950.74	46.20	17348.61	45.74	17143.63
24	45.84	15950.74	46.47	17483.31	45.53	17396.71

 https://doi.org/10.1051/e3sconf/20185102004

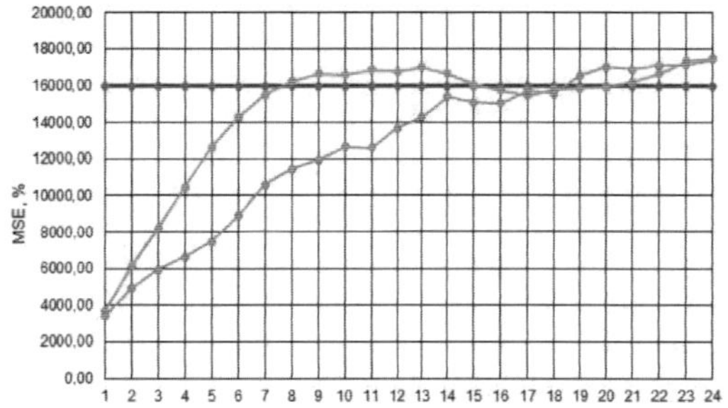

Figure 3. MSE dependence on the forecast horizon value (blue line – without history, orange line – with history, green line – history and weather)

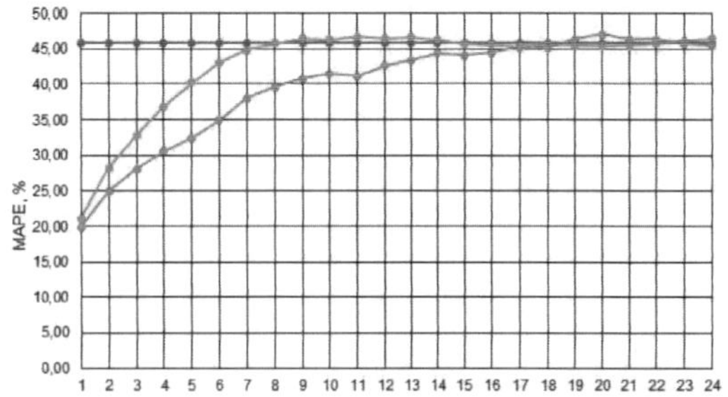

Figure 4. MAPE dependence on the forecast horizon value (blue line – without history, orange line – with history, green line – history and weather)

4 Conclusion

Within the presented research, the analysis of gradient boosting application on decision trees in the task of SPS generation operational forecasting is performed.

To improve the accuracy of generation forecasting based on the data of a local meteorological station, the authors developed a new mathematical model of operational forecasting for software implementation as part of the existing software package. Mathematical model of operational generation forecasting is implemented based on decision trees.

Within this research, test calculations of the operational generation forecast for one of the existing SPS in Russia in the form of a Python program in the Jupyter software environment were performed. In the course of the research, the accuracy of the SPS generation forecast for the operational forecast model was 75-65%, which allowed increasing the forecast accuracy by 20% in comparison with similar approaches to the SPS generation forecasting.

References

1. V. Margaret, J. Jose. Exponential Smoothing Models for Prediction of Solar Irradiance. International Journal of Advanced Research in Electrical, Electronics and Instrumentation Engineering (An ISO 3297: 2007 Certified Organization) Vol. 4, Issue 2, February 2015.
2. D.A. Snegirev, S.A. Eroshenko, R.T. Valiev, A.I. Khalyasmaa. Algorithmic realization of short-term solar power plant output forecasting. Proceedings of 2017 IEEE 2nd International Conference on Control in Technical Systems, CTS 2017. Pp. 228-231.
3. C. Severiano, F. G. Guimarães and M. W. Cohen, "Very short-term solar forecasting using multi-agent system based on Extreme Learning Machines and data clustering," 2016 IEEE Symposium Series on Computational Intelligence (SSCI), Athens, 2016, pp. 1-8.
4. C.A. Severiano, P. C. L. Silva, H. J. Sadaei and F. G. Guimarães, "Very short-term solar forecasting using fuzzy time series," 2017 IEEE International Conference on Fuzzy Systems (FUZZ-IEEE), Naples, 2017, pp. 1-6.
5. A.D. Orjuela-Cañón, J. Hernández and C. R. Rivero, "Very short term forecasting in global solar irradiance using linear and nonlinear models," IEEE Workshop on Power Electronics and Power Quality Applications (PEPQA), Bogota, 2017, pp. 1 -5.
6. S. Gupta, N. A. Shrivastava, A. Khosravi and B. K. Panigrahi, "Wind ramp event prediction with parallelized gradient boosted regression trees," 2016 International Joint Conference on Neural Networks (IJCNN), Vancouver, BC, 2016, pp. 5296-5301.
7. W. Glassley, J. Kleissl. Current state of the art in solar forecasting. California Renewable Energy Collaborative Final Report. 2013.
8. V. Prema. K. Uma Rao. Development of statistical time series models for solar power prediction // Renewable energy. 2015.

https://doi.org/10.1051/e3sconf/20185103001

The Development of Swirling Decaying Laminar Flow in an Annular Pipe

Baiman Chen[1*], Frank G.F. Qin[1], Youyuan Shao[1], Hanmin Xiao[1], Simin Huang[1] and Kelvin Ho[2]

[1]Key Laboratory of Distributed Energy Systems of Guangdong Province, College of Chemical Engineering and Energy Technology, Dongguan University of Technology, Dongguan 523808, China
[2]The University of Nottingham, Faculty of Engineering, Jalan Broga, 43500, Semenyih, Selangor Darul Ehsan, Malaysia

Abstract. Work on the hydrodynamic entry length of pipe and duct flow has been well studied over the years. The assumption of fully developed flows is commonly used in many practical engineering applications (e.g. Moody's chart). For laminar axial pipe flow, the hydrodynamic entry length can be found through the monomial proposed by Kays, Shah and Bhatti (KSB) (Lh=0.056ReDh). In contrast, several approximations exist for fully turbulent flows (i.e. 10Dh-150Dh). Through theoretical and numerical investigations, the hydrodynamic entry length for swirling decaying pipe flow in the laminar regime is investigated. It was found that, the development length Lh for the axial velocity profile changes when a tangential component is added to the mean flow. The reduction in the hydrodynamic length was found to be dependent on the inlet swirl angle θ. The results indicate that a modification can be made on the KSB equation for two-dimensional swirling annular pipe flow.

1 Introduction

The entry length of axial pipe flow has been well studied over the years [1–3]. The predictability of the entry length in pipes has benefited many practical engineering applications e.g. the estimation of wall friction factor in the fully developed region through the Moody Diagram [4], heat transfer for fully developed pipe flows [5] etc. The entry length for pipe flow is characterized by the boundary layer and the irrational core region. For the case of laminar axial flow, the velocity profile attains a fully developed state $(\partial u(r,x))/\partial x=0$, at Lh, the hydrodynamic entry length. However useful the said equations may be, it has been limited to axial flow pipe flows. It would be an interesting attempt to adapt such equations to two-dimensional flow, namely swirling/vortex pipe flow.

Swirling flow in pipes and ducts are found in many engineering applications [6, 7] (e.g. swirl tubes, vortex tubes, hydrocyclone, swirl combustors, pipe lines, mixers, flow guiders etc.). The work on various passive swirlers can be found in the wide literature [8–12].Kelvin et. al. [13–15] investigated the flow dynamics is swirling decaying flows in an annular pipe using both experimental and numerical techniques. The use of axial swirl vanes produced decaying swirl flows that can be defined using the swirl number S as functions of Re, pipe length and the vane angle. The focus of this work is to apply the numerical model developed in [7, 15] to investigate the effect of swirling flow towards the hydrodynamic entry length Lh inside and annular pipe.

Nomenclature

A	Area, m^2	u,v,w	Velocity, $m\cdot s^{-1}$
A*	Aspect ratio, A*=z/Dh	V	Volume, m^3
D	Diameter, mm	α	Thermal diffusivity of solid, m^2s^{-1}
e	Diametric clearance(Dh/2), mm	ε	Turbulence dissipation rate, m^2s^{-3}
L	Length, mm	μ	Dynamic viscosity, kg $m^{-1}s^{-1}$
P	Pressure, Pa	ρ	Density, kgm^{-3}
R, r	Radius , mm	θ	Vane angle
S	Swirl number, S=v/w	ω	Rotational speed, s^{-1}
t	Time, s		

* Corresponding author: chenbm@dgut.edu.cn

2 Methods

The experimental setup in Kelvin et. al. [15] for decaying flow is modelled using the commercial codes from ANSYS CFX 12.0. The flow field for different vanes was investigated by Baiman et. al. [14]. The focus of this work is to apply the same numerical model to investigate the effect of swirling flow towards the hydrodynamic entry length, Lh. In-depth information on the simulation model can be found in [7,13–15]. The equations (eq. 1-2) were solved for three dimensional laminar flows (μeff = μ; μt=0).

$$\nabla \cdot (\rho_f U) = 0 \tag{1}$$

The simplified time-independent non-viscous Navier-Stokes equation:

$$(\nabla \cdot U)\rho_f U - \mu_{eff}\nabla^2 U + \nabla P - \nabla \cdot (\mu_{eff}\nabla U)^T = 0 \tag{2}$$

Where

$$\mu_{eff} = \mu + \mu_t$$

The fluid domain of the annular pipe and inlet swirler was modelled (figure 1) and discretised using semi structured mesh [7].Water (ρ =1000 kg/m3; μ=0.001 Pa s) was used as the working fluid of the system. The annular tube has a length of 1000 mm (Di=30mm; Do=46mm; Dh=16 mm).Each of the vane (30°, 40° and 60°) was investigated using an average of 1x106 nodes.

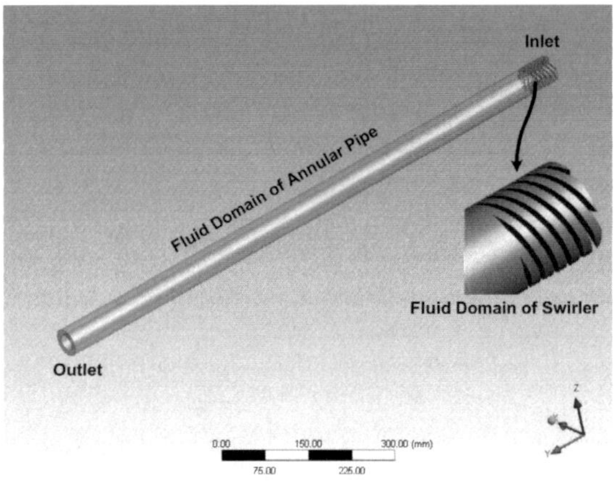

Figure 1. Computational Domain.

3 Numerical Results

The onset of this work was to investigate the effect of two-dimensional flow towards the hydrodynamic entry length. The numerical predictions for all three vane angles (30°, 45° and 60°) are shown in figure 2(a)-(f) along the annular pipe (A*=1-52) and for two different Reynolds number (Re=347 & 694). The bias of the velocity profile towards the outer pipe is caused by the swirling momentum and reduces as the vane angle increases. The symmetry of the velocity profile increase as the Swirl number decreases [14] and attains a fully developed state when A* reaches a certain length. The value of A* can then be used to estimate the hydrodynamic entry length, Lh.

From the work of Kays [1] and Shah and Bhatti [2] (KSB), it is known that the hydrodynamic entry length (Lh) can be found using the following equation.

$$\frac{L_h}{D_h} = 0.056\,\mathrm{Re} \quad \text{; For laminar internal pipe flow} \tag{3}$$

$$L_h = f(\frac{D_h}{2}, \mu, \rho, U, V, \theta) \tag{4}$$

Using the Buckingham π-Theorem eq.4 can be refined to eq.5

$$2\frac{L_h}{D_h} = f(\frac{\rho U D_h}{\mu}, Sin\theta) \tag{5}$$

The hydrodynamic entry length can be normalized using the hydrodynamic diameter of the annulus (A*=Lh/Dh).

The Reynolds number is calculated using axial velocity U.For swirling decaying flow the ratio between U and V can be related through the angle θ. [14].Dh/2 is used for an annular pipe because a developed velocity profile will only occupy Dh/2 in contrast to Dh for pipe.

$$\frac{L_h}{D_h} = \frac{A}{2} \cdot \mathrm{Re}^{\alpha} \cdot Sin\theta^{\beta} \tag{6}$$

Assuming that α ≈ 1 [1,2], the Reynolds number will have the same power as eq. 3. From the work of Baiman et. al. [7], the Resultant Reynolds number can be defined as Re/sinθ (β ≈ -1).

$$\frac{L_h}{D_h} = 0.028\frac{\mathrm{Re}}{Sin\theta} \tag{7}$$

Equation 6 shows that, the hydrodynamic entry length for swirling decaying flow in an annular duct can be obtained through a modified version of the monomial proposed by Kays, Shah and Bhatti [1,2]. A comparison between the Lh predicted using the numerical model and the correlation in eq.6 is shown in figure 3. The comparison indicates that the hydrodynamic entry length of a two-dimensional swirling decaying flow in an annular pipe is a function of the Reynolds number and the vane angle, θ.

4 Conclusion

The validated numerical model from a previous work [13] was used to study the effect of two-dimensional swirling flow towards the hydrodynamic entry length in an annular pipe. The monomial that was devised using the π-Theorem managed to predict the Lh and was comparable to the numerical predictions. However, the universality of the modified monomial and its dependency to other governing parameters (e.g. Swirl number, geometrical configurations, pressure drops etc) still requires additional work and is the subject of future work.

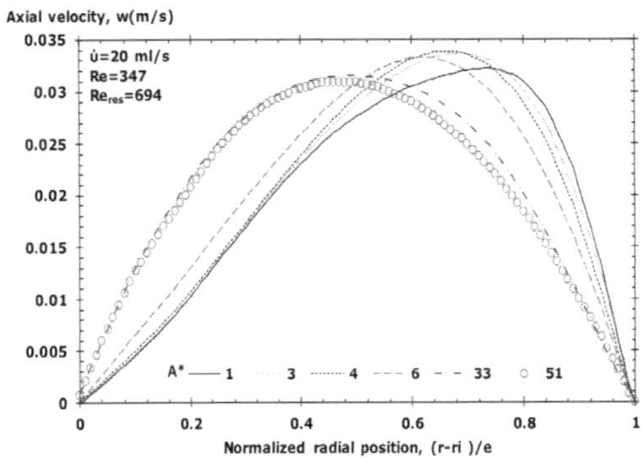

(a)

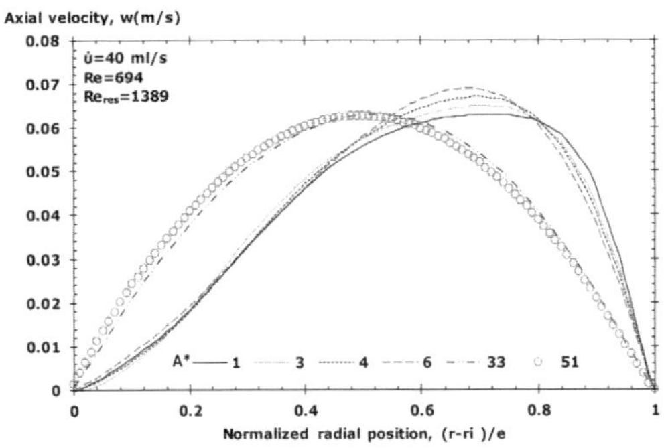

(b)

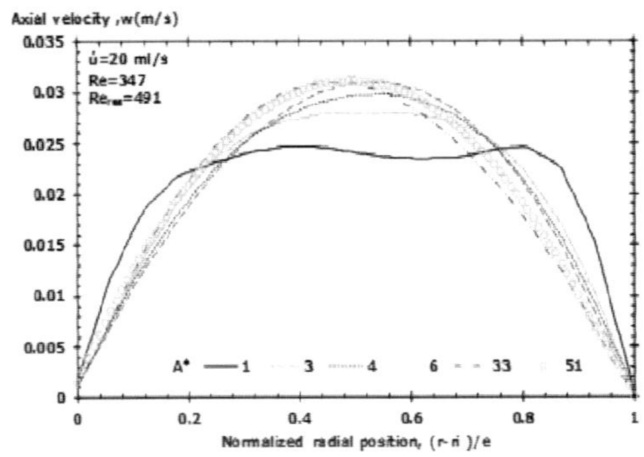

(c)

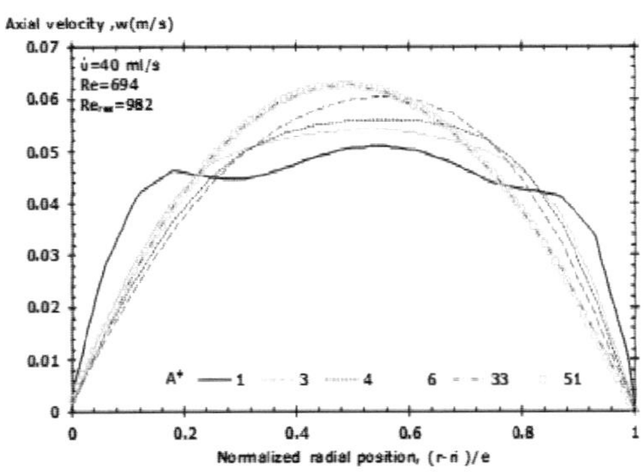

(d)

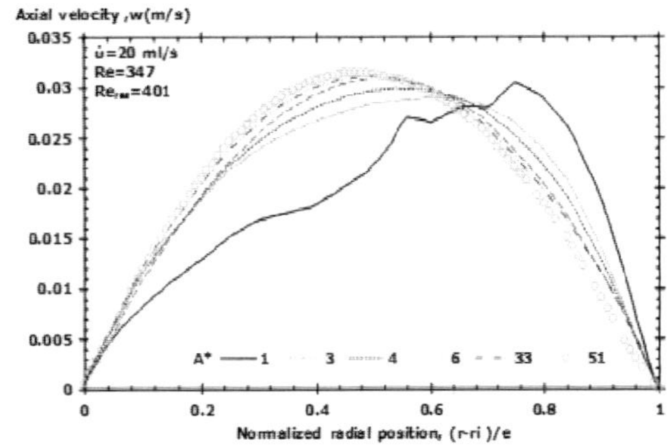

(e)

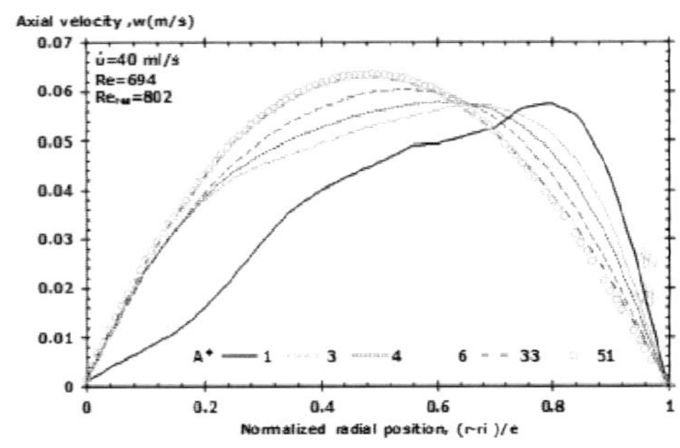

(f)

Figure 2. The development of the axial velocity profile along the annular pipe for (a-b) 30° vanes (c-d) 45° vanes (e-f) 60° vanes.

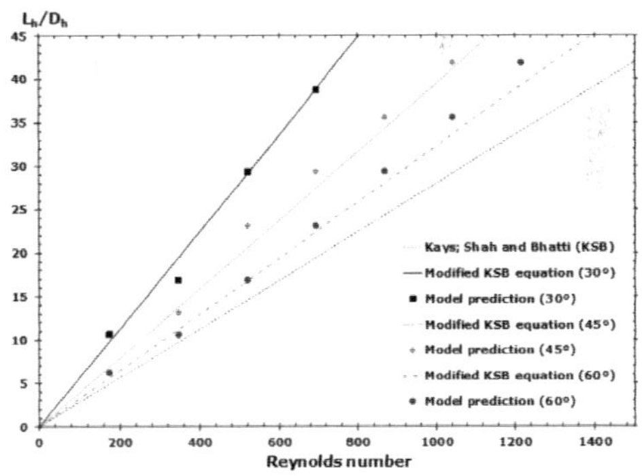

Figure 3. The hydrodynamic entry length for axial and swirling decaying pipe flow.

Acknowledgments

This work was supported by the China National Natural Science Foundation Project (No. 51406036) and Guangdong Special Innovation Project of University (No. 2016KTSCX138).

References

1. W.M. Kays, Numerical solutions for laminar-flow heat transfer in circular tubes, Trans. ASME. 77 (1955) 1265–1274.

2. R.K. Shah, M.S. Bhatti, "Laminar convective heat transfer in ducts," in Handbook of single-phase conective heat transfer, John-Wiley, 1987.
3. K. Lien, J.P. Monty, M.S. Chong, A. Ooi, The Entrance Length for Fully Developed Turbulent Channel Flow, in:15th Australas. Fluid Mech. Conf., 2004: pp. 1–4. http://sydney.edu.au/engineering/aeromech/15afmc/proceedings/papers/AFMC00093.pdf.
4. L.F. Moody, Friction factors for pipe flow, Trans ASME. 66 (1944) 671–684. doi:10.1017/S0022112004009796.
5. A. Bejan, Convection heat transfer, 3rd ed., Wiley, New Jersey, 2004.
6. R.J. Goldstein, W.E. Ibele, S. V Patankar, T.W. Simon, T.H. Kuehn, P.J. Strykowski, et al., Heat transfer--A review of 2005 literature, Int. J. Heat Mass Transf. 53 (2010) 4397–4447. http://www.sciencedirect.com/science/article/B6V3H-50GJ9DT-2/2/2826de31da348507c65d8dea83846308.
7. B. Chen, K. Ho, Y.A. Abakr, A. Chan, Fluid dynamics and heat transfer investigations of swirling decaying flow in an annular pipe Part 1: Review, problem description, verification and validation, Int. J. Heat Mass Transf. (2015). doi:10.1016/j.ijheatmasstransfer.2015.07.129.
8. R. Hreiz, C. Gentric, N. Midoux, Numerical investigation of swirling flow in cylindrical cyclones, Chem. Eng. Res. Des. 89 (2011) 2521–2539. doi:10.1016/j.cherd.2011.05.001.
9. T. Bali, B.A. Sarac, B. Ayhan, Experimental investigation of decaying swirl fl ow through a circular pipe for binary combination of vortex generators ☆, Int. Commun. Heat Mass Transf. 53 (2014) 174–179. doi:10.1016/j.icheatmasstransfer.2014.02.030.
10. M. Ahmadvand, A.F. Najafi, S. Shahidinejad, An experimental study and CFD analysis towards heat transfer and fluid flow characteristics of decaying swirl pipe flow generated by axial vanes, Meccanica. 45 (2010) 111–129. doi:10.1007/s11012-009-9228-9.
11. C. Biegger, C. Sotgiu, B. Weigand, Numerical investigation of flow and heat transfer in a swirl tube, Int. J. Therm. Sci. 96 (2015) 319–330. doi:10.1016/j.ijthermalsci.2014.12.001.
12. C. Biegger, B. Weigand, Detached Eddy Simulation of Flow and Heat Transfer in Swirl Tubes, in: High Perform. Comput. Sci. Eng. ´15, Springer International Publishing, Cham, 2016: pp. 449–461. doi:10.1007/978-3-319-24633-8_29.
13. B. Chen, K. Ho, F.G.F. Qin, R. Jiang, Y.A. Abakr, A. Chan, Validation and Visualization of Decaying Vortex Flow in an Annulus, 7th Int. Conf. Appl. Energy – ICAE2015. 75 (2015) 3098–3104. doi:10.1016/j.egypro.2015.07.640.
14. B. Chen, K. Ho, Y.A. Abakr, A. Chan, Fluid dynamics and heat transfer investigations of swirling decaying flow in an annular pipe Part 2: Fluid flow, Int. J. Heat Mass Transf. (2016). doi:10.1016/j.ijheatmasstransfer.2016.01.069.
15. K. Ho, Y.A. Abakr, A. Chan, An experimental set-up for investigating swirling decaying flow in an annular pipe, Int. Commun. Heat Mass Transf. 38 (2011) 1253–1261.

Hemp is the Future of Plastics

Ali Asghar Modi[1], Rehmatullah Shahid[1], Muhammad Usman Saeed[1], Tanzila Younas[2*]

[1]Student, Szabist, Karachi and 75600, Pakistan

[2]Assistant Professor, Szabist, Karachi and 75600, Pakistan

Abstract. Plastic is the world's most adaptable material. From bikes to food wraps and from jets to pencils, you can make anything and everything from plastics. With the infinite number of uses, plastic also have some devastating impacts on our planet. Most plastics produced today are made using petroleum-based compounds that release harmful gases into the atmosphere. Waste solutions are inefficient, and harmful by-products toxic our land, water and wildlife. Yet, consider the possibility that there was a way to deal with deliver the greater part of what we utilize causes a negative greenhouse impact, is sustainable and biodegradable and has just about an indistinguishable cost to our present techniques. Meet Hemp plastic, an only plastic that's 100% biodegradable in nature if produce by using only Hemp plant. Hemp plant consumes 4 times more carbon dioxide then other plants from atmosphere. The fiber we can produce from hemp is stronger than the conventional fiber we are using these days. This paper is intended to show numerous benefits of using hemp for the manufacturing of biodegradable plastic (HEMP PLASTIC) rather than conventional plastics.

1 Introduction

Today plastic is one of the most used materials in the world but its voyage through time was not easy. Plastics are simply the carbon chained polymers and the shape of these polymers' gives plastics their plasticity, allowing them to be molded into any shape which is why they are the world's most adaptable material. The word plastic is a typical term that is utilized for some materials of a manufactured or semi-engineered nature. The term was gotten from the Greek plastikos, which signifies "fit for enlargement."[1] Another reason of the vast use of plastics is its manufacturing cost which is quite cheap when compared to the manufacturing of same stuff with different materials.

Since each photo has two sights, similarly plastics have some devastating effects on Earth, directly as well as indirectly. Majority of the plastics is being produced by using petroleum based compounds which releases harmful gases during synthesizing, contaminating the Earth's atmosphere.

Not only the manufacturing of plastics, but also its endurance towards the degradation is also a major drawback. Every piece of plastic that was ever produced and was not recycled still exists except biodegradable plastics, which are degradable to some extent. [2]

Therefore, to reduce the increasing environmental pollution from synthetic plastic an alternative must be developed. This need can be fulfilled by use of hemp plastic, which is 100% biodegradable if produced using only Hemp plant.

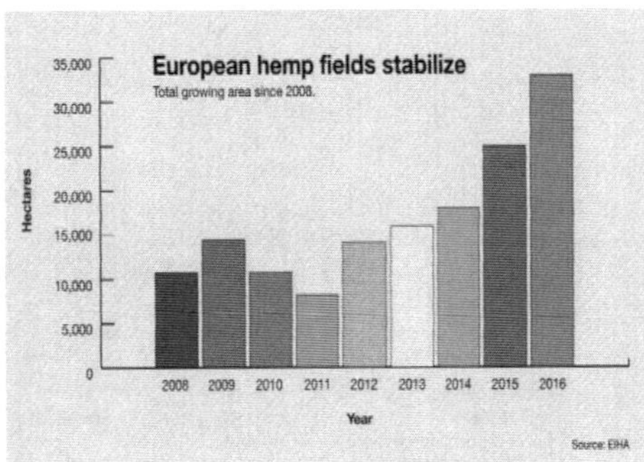

Figure 1. Increase in the fields of Hemp plant [3].

2 Types of Plastics

In order to understand the different types of plastics, one needs to understand its composition and categorizing criteria. Plastic is any synthetic or semi-synthetic organic polymer. While plastics might be produced using pretty much any natural polymer, most mechanical plastic is produced using petrochemicals. The expression "plastic" refers to the property of flexibility, which is the capacity to disfigure without breaking. The polymer used to make a plastic is quite often blended with added substances, including colorants, plasticizers, stabilizers, fillers, and fortifications. These added substances influence the concoction structure, synthetic properties, and mechanical properties of a plastic and furthermore influence its cost. The composition and the manufacturing of plastic categorize plastics mainly in Biodegradable plastics and Non-biodegradable plastics.

*Corresponding author: tanzila@szabist.eu.pk

https://doi.org/10.1051/e3sconf/20185103002

As the name suggest, biodegradable plastics are produced using petro-chemicals just like conventional plastics except some additives are included in their manufacturing which helps them in there fast degradation. Biodegradable plastics don't always break down into harmless substances, sometimes they leave behind a toxic residue and that makes them generally (but not always) unsuitable for composting. They can be breakdown using microorganisms into water, carbon dioxide and some bio chemical. And when they are put into the landfills they produce harmful greenhouse gases when breaking down, which makes them more toxic then other conventional plastics. Some examples of biodegradable plastics are polybutyrate (PBAT), polycaprolactone (PCL), polylactic acid (PLA) and polyhydroxalkanoate (PHA).

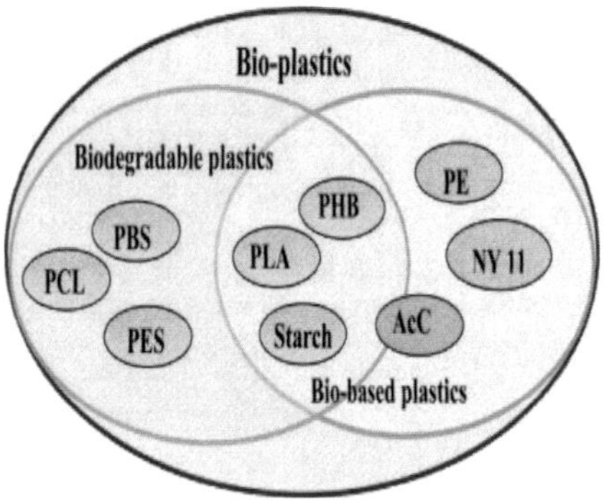

Figure 2. Examples of biodegradable plastics and bio-plastics[4].

Figure 3. Basic comparison of bio-plastic and conventional plastic [6].

Conventional plastics or non-biodegradable plastics are mostly chemically inert i.e. resistant to degradation, consequently ends up disrupting the wildlife, oceans and land. [5] And one of the most destructive impact of conventional plastic is that it is produced by the non-renewable source i.e. Petroleum.

These impacts are fueling the international drives for the production of biodegradable polymers.

3 Uses of Plastics

The flexibility of plastic materials originates from the capacity to mold, laminate or shape them, and to modify them physically and synthetically. Once anything made of plastic was considered to be of low quality but now that is history. Now days from the zip of your pants to the bikes you are riding are made up of plastics. Even the food industry uses an enormous amount of plastic. Yoghurt and butter comes in plastic tubs, Cheese in plastic wrap, water and milk in plastic containers. It is even use in construction industry in great amount. They have great versatility and combine excellent strength to weight ratio, durability, cost effectiveness, low maintenance and corrosion resistance, which make plastics an economically attractive choice throughout the construction sector. Its versatility is the reason that a Plastic is suitable for any situation. The following paragraphs will brief about the usage of plastic across the globe.

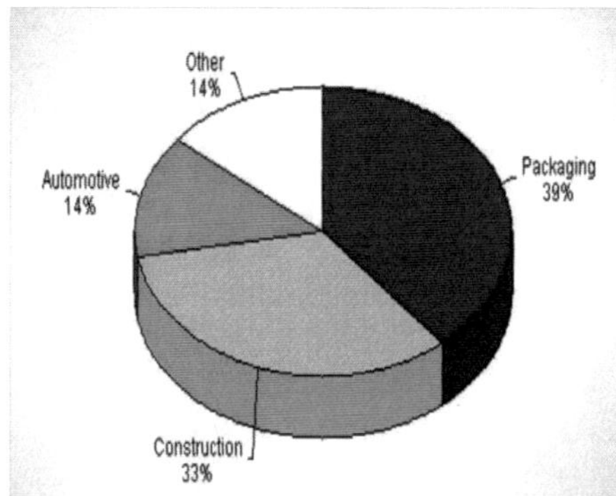

Figure 4. Uses of Plastic in different type of industries [7].

3.1. In Food Packaging

From airtight wraps to shelf stable bottles and vessels, plastic packaging plays a key role in keeping the food safe while supplying it from farm to the consumer and is a best used material for freezing foods for longer term storage. [8] Plastics are versatile, hygienic, lightweight, flexible and highly durable. Due to such qualities, plastic is used in numerous packaging applications.

Figure 5. Use of plastic in food packaging.

3.2. In Construction

Plastics are used in a growing range of applications in the construction industry. It's durability is one of the reason why it is used so frequently in construction as being corrosion resistant, allowing it to survive in any weather.[9] These days, cost is an important factor in industries and Plastics are one of the most economical material. And an amazing property of plastic .i.e. recycling beat metals in most of the places in industries.

3.3. In Transport

The financially savvy and safe transportation of individuals and merchandise is essential to our economy, cutting the weight of cars, airplanes, boats and trains can cut fuel consumption dramatically. The lightness of plastics therefore makes them priceless to the transport industry. Plastics are even used in combination with other materials are used as structural elements in all kinds of transport, even skateboards, roller blades and bicycles. [10]

Figure 6. Use of plastic in construction.

4 Comparison of Hemp with Conventional plastics

The main factors which requires the replacement of conventional plastics by Hemp plastic includes its toxicity, chemically inertness, manufacturing hazards and the limited amount of natural resources. The following paragraphs enlighten some details on the comparison of conventional and hemp plastics and the need to replace them.

4.1. Composition

The ENE compounds, Toluene, benzene, things like that, which are the most poisonous derivatives of plastics that are produced from hydrocarbons aren't found in Hemp. [11]

4.2. Manufacturing

Most of the conventional plastic today is manufactured using petroleum based compounds and petroleum consumption is considered harmful to the environment, given the measure of waste created amid the refinement procedure and the destructive techniques used to separate oil from the earth. And simultaneously we are losing one of most important natural resource i.e. Petroleum in the manufacturing of plastics. On the other hand, hemp plastic is solely produced using the cellulose extracted from hemp plant and therefore causes no toxicity during its production.

4.3. Biodegradable and Recyclable

Hemp plastic is 100% biodegradable and recyclable when it is made using completely a Hemp plant. We traditionally think of recycling as the conversion of waste into usable material, although if we use more hemp plastic, we could introduce a new cycle that would be much more environmentally friendly. Plants are harvested and broken down into vital components for production and a product is produced. After its utilization, it is put into a landfill where nature will run its course and separate it into required supplements basic for the plant's development and the cycle is finished. [12]

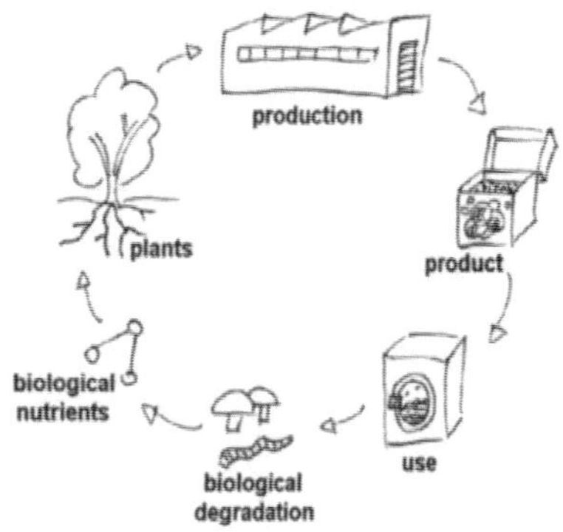

Figure 7. Cycle of hemp plastic.

4.4. Environment friendly

Hemp absorbs four times the amount of carbon dioxide as trees do amid its snappy 12-14 week develop cycle. Delivering hemp plastic likewise requires 22-45% less energy than non-renewable energy source based items [13]. We can have fields, acres and acres, hectares of hemp farms that are pulling carbon dioxide out of the atmosphere (as plants do). Then, that carbon from the carbon dioxide in the atmosphere gets used to make the plastics, and the plastics, when they are going into a landfill and they are no longer usable, will biodegrade bringing carbon back into the soil. So, it's essentially carbon negative, pulling carbon out of the atmosphere and putting it back into the soil.

4.5. Applications

The strength of hemp plastic makes it appealing in several industries. The automotive, building, and packaging industries are all being turned on to hemp plastic. It is also lightweight and has a very high ratio of density to weight. This allows it to potentially be used in aerospace to save weight on heavy structures. Due its versatility, hemp can be used anywhere for any purpose replacing conventional plastics.

4.6. Production cost

Plastics produced using petroleum compounds are quite cheap and easy to manufacture and on the other hand, the processes involve in the production of biodegradable plastic or specifically hemp plastics are quite expensive and use more effort than the production of conventional plastics.

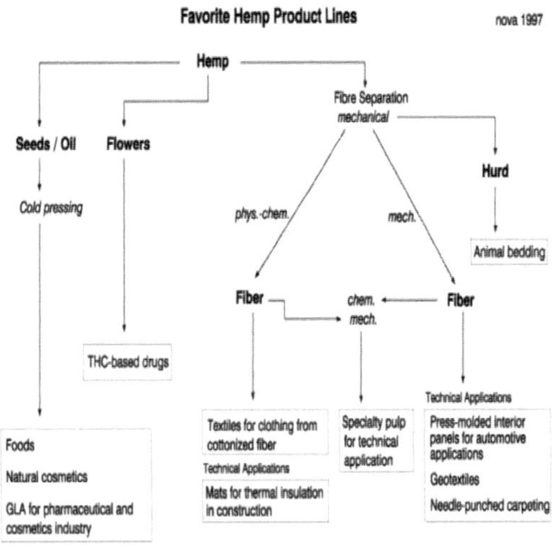

Figure 8. Applications of Hemp plant [14].

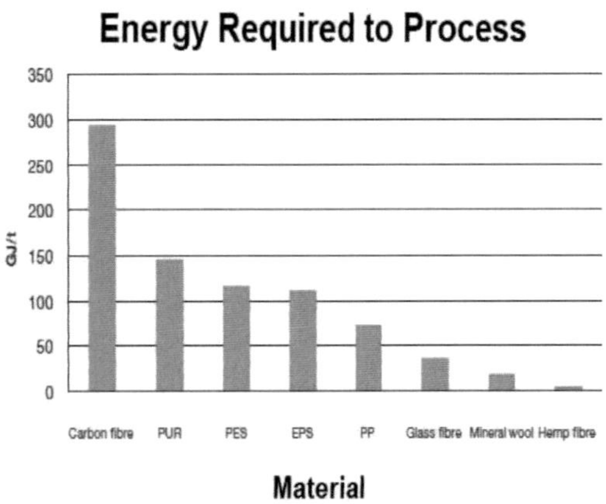

Figure 9. Energy required for the production of different type of plastics.

4.7.Toughness and Flexibility

Hemp Plastics can be five times stiffer and 2.5 times stronger than polypropylene, and it will not cause wear and tear to the screw and the mold like glass fibers do, and unlike glass fibers, it does not pose safety and health risks. [14] Its flexibility is one of the greatest factors for its superiority over conventional plastics.

5 Conclusion

With an ever increasing number of plastics being utilized in human lives and expanding weight being set on limits accessible for plastic wastes disposal, the need for biodegradable plastics and biodegradation of plastic wastes have gained increasing importance in the last few years. Also the available sources of petroleum oil is declining (which is the major source of conventional plastics) adds up for the need of an alternative way of creating plastics. In any case, countries all through the world do understand that petrochemical utilization should be lessened, and hemp is being perceived as a reasonable contrasting option to these plastics. There are number of different bio plastics are available but the best one among them is Hemp. Many factors as discussed above prove hemp to be better than other bio plastics such as its lightweight, versatility, degradation time etc. Though producing plastics using only hemp plants causes more efforts and money but the impacts that conventional and other bio plastics are causing to this Earth are getting unbearable and therefore conventional plastics should be completely replaced with hemp plastics.

References

1. Precision Engineered Products [Online]. Available: http://www.pepctplastics.com/resources/connecticut-plastics-learning-center/an-introduction-to-plastics/
2. L.Concialdi, May 03,2016. [Online]. Available: https://discovercbd.com/blogs/cbd-news/105784134-benefits-of-hemp- plastic-over-traditional-plastic
3. Hemp Today. March 27,2017. [Online]. Available: https://hemptoday.net/european-hemp-fields-grow/
4. L.Y. Beeker, M. Pringle Joshua, M. Pearce. International Journal of Molecular Sciences (September, 2009)
5. A.M.Helmenstine, November 15, 2016. [Online]. Available: https://www.thoughtco.com/plastic-chemical-composition-608930
6. October, 2012. [Online]. Available: https://www.slideshare.net/ShwetaChand2/bioplastic-66639489
7. Presto Testing Instruments. [Online]. Available: http://www.testing-instruments.com/blog/problems-faced-by-the-manufacturers-in-plastic-industry-and-their-solutions/
8. America's Plastic Makers. [Online]. Available: https://www.plasticpackagingfacts.org/plastic-packaging/plastic-packaging-by-industry/food-beverage/
9. British Plastic Federation. 2018 [Online] Available: http://www.bpf.co.uk/plastipedia/applications/default.aspx
10. T. Johnson. April 01, 2017. [Online] Available: https://www.thoughtco.com/uses-of-plastics-820359
11. J.Hasse, May 05, 2017. [Online]. Available: https://www.benzinga.com/markets/emerging-markets/17/05/9405826/why-hemp-could-be-the-future-of-plastics
12. National Hemp Association (NHA), May 17, 2016. [Online]. Available: http://nationalhempassociation.org/hemp-plastic/
13. C. Woodford, June 11, 2017 [Online]. Available: http://www.explainthatstuff.com/bioplastics.html
14. M. Karus and G. Leson. International Hemp Association. Vol. 4 (June 01, 1997)

 https://doi.org/10.1051/e3sconf/20185103003

Production of Hydrogen by Steam Methane Reformation Process

T Younus[1*], A Anwer[2], Z Asim[2], M S Surahio[2]

[1]Assistant Professor, Szabist, Karachi and 75600, Pakistan
[2]Student, Szabist, Karachi and 75600, Pakistan

Abstract. Upcoming hydrogen economy is on rise on political agenda due to growing need of hydrogen. Natural occurrence of hydrogen cannot satisfy the present need of hydrogen. It produces a wide gap between current hydrogen requirement and amount of hydrogen present in earth. To counter this problem, hydrogen is produced commercially in industries through various methods. Among all these methods, SMR (Steam Methane Reforming) process is considered most feasible for being economically cheap as compared to other methods. Being economical does not necessarily mean being eco-friendly. Industrialist does not switch on alternative methods and continue using SMR process which is producing a devastating impact on atmosphere by increasing the amount of CO2 (carbon dioxide). Greenhouse effect of carbon dioxide makes it one of the primary sources of increasing global warming in earth's atmosphere. Apart of other uses, Hydrogen can also be used as eco-friendly energy source as compared to fossil fuel used as energy source. In this paper, the procedure of production of hydrogen through SMR process is reviewed in detail and its pros and cons are discussed.

1 Introduction

Currently electricity, natural gas, solar heat, gasoline and diesel serve as energy source to drive this world [1, 2]. These energy carriers are converted from primary energy sources such as coal, petroleum, underground methane and nuclear energy. Conversion of primary source to usable source produces harmful by-products which have made necessary for humans to convert to more eco-friendly and clean energy. These energy sources include wind energy, solar energy, tidal energy and hydrogen etc. hydrogen being limited in its natural form need to be produced commercially. Once produced can serve as one of the clean eco-friendly energy source with no carbon emission. Currently hydrogen is mainly used to produce ammonia and methanol. It is also used for hydro treating refineries. Hydrogen is widely used for hydrogenation reactions for unsaturated hydrocarbons, aromatics, aldehydes and ketones and nitrogen compounds. Other uses include use of hydrogen in food industry, semiconductor industry and metallurgical industry. Some more uses include mixture of hydrogen and nitrogen in synthesis of ammonia and mixture of hydrogen and carbon oxide in synthesis of methanol, liquid hydrocarbons and higher alcohols [3]. Taking all these uses of hydrogen in account, some methods for commercial production of hydrogen have been proposed having their own advantages and disadvantages.

2 Literature Review

Extensive usage of fossil fuel results in the increased amount of greenhouse gases and other air pollutants, this highlights the need of new methods for the production of energy. Developing hydrogen as an energy carrier may answer to this open challenge [4-6]. Bio-Fermentation is one of the technique to produce hydrogen [7]. Hydrogen can be a sustainable source to meet the- need of energy in the near future is discussed by Turner [8]. Hydrogen is sustainable source of renewable energy [9, 10]. Geological hydrogen storage methods and its transmission lines for transportation is discussed by Leighty [11]. Hydrogen is found to be one of the main energy carrier for the production of sustainable energy [12 - 17]. Economy may face some positive prospective along with few challenges which is in detailed discussed by Abbasi and Abbaso [18]. Residual gasses production in greenhouse is greatly reduced by using hydrogen as compared to fossil fuels. A new era of clean energy is emerging nowadays, in which hydrogen plays a prominent role [19-25]. The environmental effect of hydrogen production can be assessed through a life cycle of system used presented by Smitkova [26]. Hydrogen is not freely available in atmosphere but can be extracted/produced using several systems. Coal, water, natural gas, and oil are the sources that produce hydrogen in its purest form. Among many processes few are; (a) Steam Methane Reforming (SMR), (b) Coal Gasification, (c) Water Electrolysis, (d) Fermentation of Biomass and (e) Thermochemical Cycles [27-34]. More than 90% of hydrogen is produced using SMR, Dufour et al. [35] stated that it is one of the main cause of releasing carbon dioxide. However hydrogen can also be produced through separation of water molecules using electrical

*Corresponding author: tanzila@szabist.edu.pk

power (electrolyzers). In order to produce 1,000 Joules of energy 1,600 Joules of hydrogen is required [36].

Techniques that can be used to produce hydrogen can be categorized according to the source used for production of hydrogen. Among all these techniques, SMR (steam methane reforming) method is considered to be the most efficient and feasible procedure. [Being economical and producing 99.9% pure hydrogen, a lot of industrialists are investing on commercial production of hydrogen through Steam Methane Reforming Process]. The steam reforming procedure seems to be a simple procedure as product composition can be expressed by simple thermodynamics but in reality it is a complex procedure which includes a combination of complex mechanical design, catalyst and heat transfer.

In past years, there have been modifications in this process which allowed more efficient and less expensive commercial units due to the changes in material of reformer tubes, better control and understanding of carbon limits and better catalyst and process concept developed over time.

3 Methodology

A simple hydrogen plant producing hydrogen by SMR process is explained below. Natural gas is heated beforehand in the waste side of the reformer and zinc oxide catalyst removes sulphur. Treated steam is included, and the combination of natural gas and steam is further heated before going in the tabular reformer. In this step, transformation of equilibrium of hydrogen, carbon monoxide and carbon dioxide occurs in presence of nickel based catalyst. The gas leaves the reformer and is cooled by production of steam before going in the shift converter. In presence of shift catalyst, further production of hydrogen takes place by processing carbon monoxide and steam into carbon dioxide and hydrogen. The converted gas is cooled to room temperature before going into PSA unit. Pure hydrogen gas is produced and the off-gas from PSA unit is used in the reformer as fuel supplemented with natural gas [37].

Figure 1 shows the summary of steam methane reformation process. Whole process of SMR (steam methane reforming) process can be summarized in three equations. These are

$$CH_4 + H_2O = 3H_2 + CO \ (\Delta H298K = 206\ kJ/mol) \quad [38] \quad (1)$$

$$CO + H_2O = CO_2 + H_2 \ (\Delta H298K = -41\ kJ/mol) \quad [38] \quad (2)$$

CO_2 is also produced as a by-product when hydrogen is produced [2]. Quantity of CO_2 produced mainly depends on the presence of carbon element and technique used for production [1]. Chemical process of carbon produces 1 volume of CO_2 per volume of H_2. Whereas 0.25 volume of CO_2 per volume of H_2 is produced by steam reforming of methane and higher efficiency plants produces 0.37 volume of CO_2 per volume of H_2. It can also be represented as 8.1 tone CO_2 per ton H_2 [39]. These are pretty high values of carbon emission and when released in atmosphere can produce pretty harmful effects on atmosphere producing greenhouse effects and increasing average temperature of earth. Carbon emission is one of the major reasons for global warming causing climatic change season cycle all around the world. To counter these effects, a technique termed as carbon sequestration is proposed.

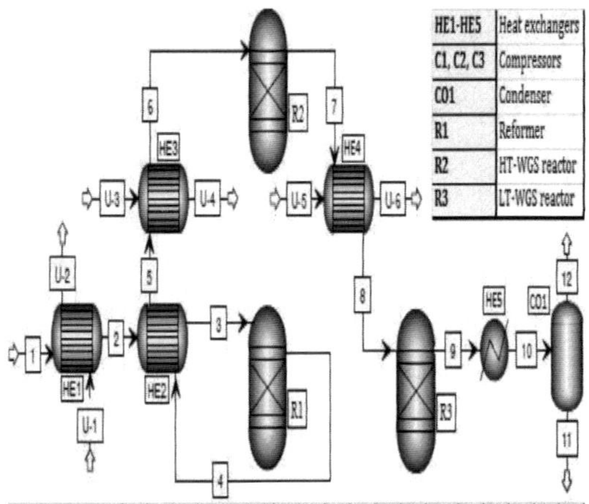

Streams		1	2	3	4	5	6	7	8	9	10	11	12	U-1	U-2	U-3	U-4	U-5	U-6
Gas substream	CH_4	√	√	√	√	√	√	√	√	√	√		√						
	H_2O	√	√	√	√	√	√	√	√	√	√	√				√	√	√	√
	CO				√	√	√	√	√	√	√		√						
	CO_2				√	√	√	√	√	√	√		√						
	H_2				√	√	√	√	√	√	√		√						
	Flue gases													√	√				
	T(°C)	25	208	490	850	614	400	400	210	210	110	25	25	1000	245	101	224	101	224

Figure 1. Steam Methane Reformation Process Working Principle [38]

4 Dumping of Carbon through Carbon sequestration process

Carbon sequestration is the long-term storage of carbon in oceans, soils, vegetation (especially forests), and geologic formations [4]. It is still under experimentation phase. According to this concept, carbon dioxide will be mainly dumped into deep ground and under deep sea beds. Apparently it appears to be a good idea to get rid of excessive carbon dioxide from atmosphere but it has its own drawbacks that cannot be ignored. Some of its major drawbacks are listed below.

The most substantial risk associated with CCS is the leakage of CO2 from storage sites. There is little encounter with geological deposition of natural gas and CO2 for time interval of almost 10-20 years but a long term deposition of hundreds or thousands of years is not proved[40]. Excessive concentration of carbon dioxide can change the carbon content of soil. It is good for some crops that can grow well in excessive but overall effect can be dangerous. Most of the crops need limited proportion of carbon content and increased carbon concentration can damage crop's growth and change crop cycle as well.

Another option of storing excessive carbon dioxide is to store in deep sea beds. Excessive carbon can play a major role in growth of planktons and reefs but overall marine could get in danger thus disturbing the marine life cycle and ultimately food chain.

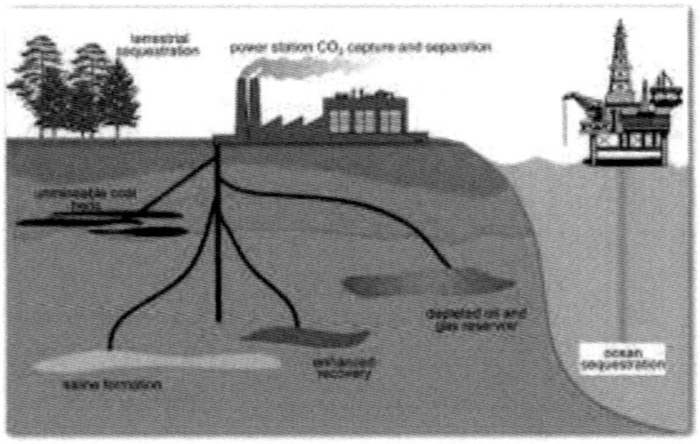

Figure 2. Carbon Sequestration Methods [41]

Fig. 2 shows different methods of carbon sequestration according to the application. Land industries are dumping carbon underground and ocean industries are dumping carbon in ocean beds.

5 Alternate Method

Alternate methods for producing carbon emission free hydrogen is following bio- hydrogen route.
Bio -hydrogen is fuel obtained by microbial metabolism, which is bioethanol or biogas. Bio hydrogen can be produced by mainly two mechanisms. These mechanisms include sunlight based microbial photosynthesis using water or organic subtract and dark fermentation by heterotrophic bacteria using sugar, starches etc. difference between these two methods is based on microorganism used. Sunlight based micro- organic photosynthetic uses green microalgae, cyano-bacteria whereas dark fermentation uses fermentative bacteria. However bio-hydrogen produces about 20-25% of hydrogen [42].

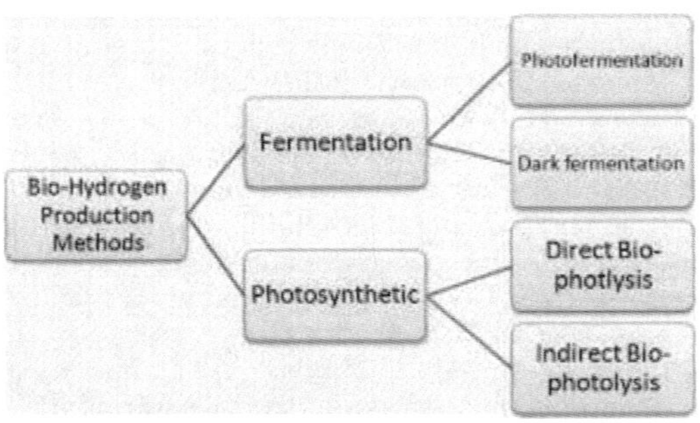

Figure 3. Bio Hydrogen Production Methods [26-33]

Fig. 3 shows the different type of methods of bio hydrogen methods of production.

6 Result

Analytically bio hydrogen procedure for producing commercial hydrogen appears to be costly and less efficient as compared to the SMR (steam methane reforming) process. Steam methane reforming process is more efficient as compared to other methods of production of Hydrogen. Huge amount of carbon dioxide emission from steam methane reforming process raises questions on future of hydrogen as alternative energy source.

7 Discussion

Production of hydrogen through steam methane process produces about 99.9% pure hydrogen which can be used as alternative energy source but produces huge amount of carbon dioxide which will affect earth's atmosphere indirectly. On both conditions i.e. producing hydrogen from steam methane process and using it as an alternate energy source or directly obtaining energy by burning fossil fuel have adverse effects on earth's atmosphere and increasing global warming rapidly.

8. Conclusion

This paper gives deep insight about production of Hydrogen through Steam Methane Reformation process and why it should not be continued in future. Carbon emission and its effect on atmosphere is one of the major issue which needs to be controlled on right time. Not taking any action on right time can cause a huge setback to the humanity by in forms of natural catastrophes.

References

1. Koroneos, C, Spachos, T, Moussiopoulos, N, Exergy analysis of renewable energy sources, Renewable Energy, 2003; Volume 28, Issue 2, 295-310
2. Younas, T, Saadat, M, Hussain, I, Rashied, Z, "Design and Fabrication of Power Generator through Busy Roads", Energy Procedia, 2017; 118C, 104-109: 1876-6102
3. Antzaraa, A., Heracleousa, E., B. Bukurc, D., & A. Lemonidoua, A. (2014). Thermodynamic analysis of hydrogen production via chemical looping steam methane reforming coupled with in situ CO2 capture. Energy Procedia 63, 6576 – 6589.
4. Gadalla, MA, Ratlamwala, TAH, Dincer, I., Kanoglu, M. Performance assessment of an integrated absorption cooling-hydrogen liquefaction system using geothermal energy. International Journal of Exergy 2013; 12 (2): 205-25.
5. Clark, WW, Rifkin, J, O'Connor, T, Swisher, J, Lipman, T, Rambach, G. Hydrogen energy stations: along the roadside to the hydrogen economy. Utilities Policy 2005; 13 (1): 41-50.
6. Ratlamwala, TAH, Dincer, I. Performance assessment of solar based integrated Cu-Cl systems for hydrogen production. Solar Energy 2013; 95: 345-56.

7. Argun, H, Kargi, F. Bio-hydrogen production by different operational modes of dark and photo-fermentation: An overview. International Journal of Hydrogen Energy 2011; 36 (13): 7443-59.
8. Turner, JA. Sustainable hydrogen production. Science 2004; 305: 972-4.
9. Dincer, I., Ratlamwala, TAH. Development of Novel Renewable Energy Based Hydrogen Production Systems: A Comparative Study. Energy Conversion and Management 2013; 72: 77-87.
10. Ratlamwala, TAH, Dincer, I and Aydin, M. Energy and exergy analyses and optimization study of an integrated solar heliostat field system for hydrogen production. International Journal of Hydrogen Energy 2012; 37: 18704-12.
11. Leighty, W. Running the world on renewable: Hydrogen transmission pipelines and firming geologic storage. International Journal of Energy Research 2008; 32: 408-26.
12. Dincer, I, Ratlamwala, TAH. Importance of exergy for analysis, improvement, design, and assessment. Wiley Interdisciplinary Reviews: Energy and Environment 2013; 2 (3): 335-49.
13. Bicakova, O, Straka, P. Production of hydrogen from renewable resources and its effectiveness, International Journal of Hydrogen Energy 2012; 37: 11563-78.
14. Moriarty, P, Honnery, D. Hydrogen's role in an uncertain energy future. International Journal of Hydrogen Energy 2009; 34: 31-9.
15. Dincer, I. Environmental and sustainability aspects of hydrogen and fuel cell systems. International Journal of Energy Research 2007; 31: 29-55.
16. Ratlamwala, TAH, Dincer, I, Gadalla, MA and Kanoglu, M. Thermodynamic analysis of a new renewable energy based hybrid system for hydrogen liquefaction. International Journal of Hydrogen Energy 2012; 37 (23): 18108-17.
17. Ratlamwala, TAH, Dincer, I. Development of a geothermal based integrated system for building multigenerational needs. Energy and Buildings 2013; 62: 496-506.
18. Abbasi, T, Abbaso, SA. 'Renewable' hydrogen: Prospects and challenges. Renewable and Sustainable Energy Reviews 2011; 15: 3034-40.
19. Gadalla, MA, Ratlamwala, TAH, Dincer, I. Energy and exergy analyses of an integrated fuel cell and absorption cooling system. International Journal of Exergy 2010; 7(6): 731-54.
20. Muradov, NZ, Veziroglu, TN. "Green'' path from fossil-based to hydrogen economy: An overview of carbon-neutral technologies. International Journal of Hydrogen Energy 2008; 33: 6804-39.
21. Midilli, A, Dincer, I. Development of some exergetic parameters for PEM fuel cells for measuring environmental impact and sustainability. International Journal of Hydrogen Energy 2009; 34: 3858-72.
22. Ratlamwala, TAH, Dincer, I, Gadalla, MA. Performance analysis of a novel integrated geothermal-based system for multigeneration applications. Applied Thermal Engineering 2012; 40(1): 71-79.
23. Midilli, A, Dincer, I. Hydrogen as a renewable and sustainable solution in reducing global fossil fuel consumption. International Journal of Hydrogen Energy 2008; 33: 4209-22.
24. Mohammed, MAA, Salmiation, A, Wan-Azlina, WAKG, Amran, M, Fakru'l-Razi, A, Taufiq-Yap, YH. Hydrogen rich gas from oil palm biomass as a potential source of renewable energy in Malaysia. Renewable and Sustainable Energy Reviews 2011; 15: 1258-70.
25. Ratlamwala, TAH, Dincer, I, Gadalla, MA. Thermodynamic analysis of a novel integrated geothermal based power generation-quadruple effect absorption cooling-hydrogen liquefaction system. International Journal of Hydrogen Energy 2012; 37 (7): 5840-49.
26. Smitkova, M, Janicek, F, Riccardi, J. Life cycle analysis of processes for hydrogen production. International Journal of Hydrogen Energy 2011; 36: 7844-51.
27. Abuadala, A, Dincer, I. A review on biomass-based hydrogen production and potential applications. International Journal of Energy Research 2012; 36: 415-55.
28. Ratlamwala, TAH, Dincer, I. Comparative efficiency assessment of novel multi-flash integrated geothermal systems for power and hydrogen production. Applied Thermal Engineering 201; 48: 359-66.
29. Abd-Alla, MH, Morsy, FM, El-Enany, AE. Hydrogen production from rotten dates by sequential three stages fermentation. International Journal of Hydrogen Energy 2011; 36: 13518-27.
30. Joshi, AS, Dincer, I, Reddy, BV. Exergetic assessment of solar hydrogen production methods. International Journal of Hydrogen Energy 2010; 35: 4901-8.
31. Ratlamwala, TAH, Gadalla, MA, Dincer, I. Performance assessment of an integrated PV/T and triple effect cooling system for hydrogen and cooling production. International Journal of Hydrogen Energy 2011; 36(17): 11282-91.
32. Ratlamwala, TAH, El-Sinawi, AH, Gadalla, MA, Aidan, A. Performance analysis of a new designed PEM fuel cell. International Journal of Energy Research 2012; 36 (11): 1121-32.
33. Tsuchiya, H. Innovative renewable energy solutions for hydrogen vehicles. International Journal of Energy Research 2008; 32: 427-35.
34. Ratlamwala, TAH, Gadalla, MA, Dincer, I. Thermodynamic analysis of an Integrated PEMFC-TEARS-geothermal system for sustainable buildings. Energy and Buildings 2012; 44(1): 73-80.
35. Dufour, J, Serrano, DP, Galvez, JL, Moreno, J, Garcia, C. Life cycle assessment of processes for hydrogen production: environmental feasibility and reduction of greenhouse gases emissions. International Journal of Hydrogen Energy 2009; 34: 1370-76.

36. Burman, G. Hydrogen Isn't Yet the Miracle Fuel of the Future. The Fresno Bee 2003.
37. Hallenbeck, P., & Benemann, J. (n.d.). Biohydrogen - The Microbiological Production of Hydrogen Fuel. In P. C. Hallenbeck, & J. R. Benemann, Biotechnology – Vol.VII.
38. New York State Energy Research and Development Aut. (n.d.). Amiqweb.es. Retrieved fromamiqweb.es: www.amiqweb.es/app/download/.../6hydrogenpr oductionsteammethanereforming.pdf
39. Nielsen, J., & -Nielsen, T. (2002). Large-scale Hydrogen Production. T. Cattech, 6(4), 150-159.
40. Shahbazi, A., & Nasab, B. (2016). Carbon Capture and Storage (CCS) and its Impacts on Climate Change and Global Warming. Journal of Petroleum & Environmental Biotechnology.
41. Paper online: https://www.quora.com/How-is-carbon-sequestration-useful-to-the-environment
42. Udengaard, N. (2004). Hydrogen Production by Steam Reforming Of Hydrocarbons. Prepr. Pap.-Am. Chem. Soc., Div. Fuel Chem., 49(2), 906.

Polyaniline Modified with Cobalt-Hexacyanoferrate (PmCH) as an Adsorbent for Removal of Heavy Metals from Aqueous Solutions

Nima Moazezi[1,*] and Mohammad Ali Moosavian[1]

[1]School of Chemical Engineering, College of Engineering, University of Tehran, 11155-4563, Tehran, Iran

Abstract. In this study, polyaniline modified with cobalt-hexacyanoferrate (PmCH) composite was synthesized and characterized for removal of Rb^+, Cd^{2+}, Zn^{2+}, Pb^{2+}, and Ni^{2+} by FTIR and XRD. The effect of pH, adsorbent dosage, ionic strength, contact time, initial ion concentration, and temperature were studied. The competition adsorption experiments between metal ions were investigated. Batch desorption was also conducted to evaluate the reusability of PmCH. The maximum adsorption capacities were 96.15, 27.17, 17.85, 19.15, and 4.76 mg g^{-1} of Rb^+, Cd^{2+}, Zn^{2+}, Pb^{2+}, and Ni^{2+}, respectively. The optimum pH was determined at natural pH of each solution.

1 Introduction

As water pollutants, heavy metals have come up as an interesting area of water research [1]. One of the most popular techniques to removal of these toxic compounds is adsorption process [2]. Due to strong adsorption capability to removal of metal ions, polyaniline can be used as a good prospect for adsorption applications [3]. On the other hands, cobalt-hexacyanoferrate is considered as an effective compound in ion exchange and storage potential of counter cations [4]. In this study, polyaniline modified with cobalt-hexacyanoferrate (PmCH) composite was prepared and characterized for removal of Rb^+, Cd^{2+}, Zn^{2+}, Pb^{2+}, and Ni^{2+} from aqueous solutions.

2 Materials and Methods

2.1 Chemicals and instrumentations

$Ni(NO_3)_2$, cobalt (II) chloride hexahydrate, aniline, HCl, KOH, ammonium peroxydisulfate, $Pb(NO_3)_2$, $ZnCl_2$, $Cd(NO_3)_2$, and RbCl were used in analytical grade (Germany, Merk). FTIR (Bruker, Vector 22) and XRD (STOE, STIDY-MP, Germany) measurements were performed to detect the morphological studies of polyaniline and PmCH. The ion concentrations were analyzed using an ICP-OES (Optima 7300 DV, Perkin Elmer).

Nomenclature	
V	Solution volume
C_i / C_e	Initial / Equilibrium concentration of metal ion
W	Adsorbent weight
q_e / q_t	Adsorption capacity at equilibrium / at any time
k_1 / k_2	The rate constant of pseudo-first-order / pseudo-second-order adsorption
h	The initial adsorption rate
k_f / k_L	The Freundlich constant / Langmuir constant
n	Adsorption intensity and bond energy (Freundlich isotherm)

* Corresponding author: nmoazezi@ut.ac.ir

https://doi.org/10.1051/e3sconf/20185103004

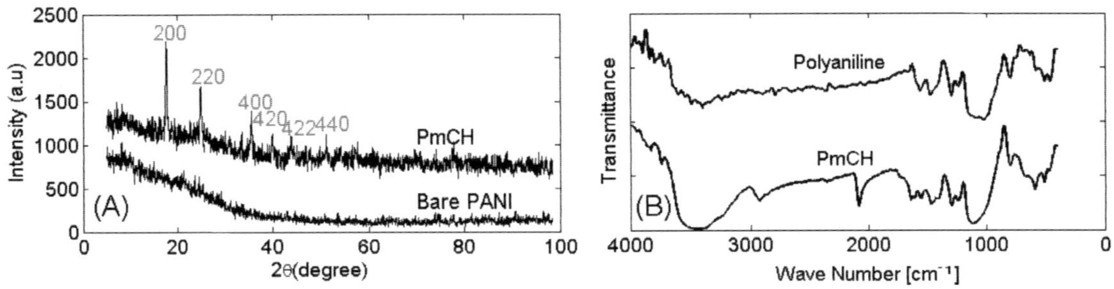

Figure 1. Characterization of polyaniline and PmCH by (A) XRD (B) FTIR.

2.2 Synthesis of Polyaniline

Ammonium peroxydisulfate (one gram) was dissolved in 100 mL of 1 mol L^{-1} HCl media. After 30 min, one millilitre of aniline was dropwise added to form a dark green solution. After 5 h of continuous stirring, the solution was filtered and its precipitate was collected. Then, the precipitate was finally washed and dried with distilled water.

2.3 Synthesis of PmCH

Cobalt (II) chloride hexahydrate (2 mol L^{-1}), potassium ferricyanide (3 mol L^{-1}), and ammonium peroxydisulfate (one gram) were dissolved in 100 mL of 1 mol L^{-1} HCl. After 30 min, one millilitre of aniline was dropwise added to form a dark green solution. After 5 h of continuous stirring, the solution was filtered and its dry precipitate was collected. Then, the precipitate was finally washed and dried with distilled water at room temperature.

2.4 Adsorption and desorption of heavy metal ions

In general, one gram of dry adsorbent was added to 100 mL of solutions containing Rb^{+}, Cd^{2+}, Zn^{2+}, Pb^{2+}, and Ni^{2+} ions. These solutions were shaken in a thermostatic shaker. The initial pH of solution for each ion was adjusted in the range of 2.5 to 9. In order to evaluate the effects of concentration of heavy metals, the initial concentration of the solutions were considered between 10-100 mg L^{-1}. The adsorbent dosage was considered at 3.5 gr L^{-1}. The adsorption capacity and ion removal were calculated by [2]:

$$q_e = \frac{C_i - C_e}{W} \times V \quad ; \quad removal(\%) = \frac{C_i - C_e}{C_i} \times 100 \qquad (1)$$

3 Equations and mathematics

3.1 Characterization of Polyaniline and PmCH

The XRD patterns of polyaniline and PmCH are depicted in figure 1A which illustrates that polyaniline and PmCH have an amorphous and the cubic structures, respectively. The reflections of PmCH at 200, 220, 400, 420, 422, and 440 were located at 18, 24.8, 35.5, 40.0, 43.9, and 51.0°, respectively (JCPDS 5-0036, 5-0037). The changes in FTIR spectra of polyaniline and PmCH at 2090 cm^{-1} (Fe-CN-Co) indicates that cobalt-hexacyanoferrate have been successfully added to the polyaniline; see figure 1B. The bands at 788, 1105, 1286, 1444, and 1610 cm^{-1} correspond to C–H bending, C–N bonds in the benzenoid rings, C–N bonds in the quinoid rings, benzenoid rings, and quinoid rings, respectively [2]. A broad bond at around 3100–3650 and 2850–2950 cm^{-1} are assigned to O–H stretching for the hydrogen bonded hydroxyl groups and the alkyl stretching groups, respectively [3].

3.2 Effect of pH

The effects of initial pH on the adsorption of each metal ion onto PmCH were evaluated at pH 2-6 and an initial ion concentration of 100 mg L^{-1}. As can be seen in figure 2A, adsorption capacity of PmCH was enhanced with increasing the pH. At low pH, the surface of PmCH has a positive charge. The negative charge density of PmCH surface was increasing with increasing the pH. At pH above the zero point charge, the functional groups of PmCH surface have the negative charge; hence, maximum cation sorption occurs at high pH [5]. At high pH, due to salt formation of Cd^{2+}, Zn^{2+}, Pb^{2+}, and Ni^{2+} some precipitations were appeared. So, these studies do not continued more than precipitations pH. Fortunately, the optimum pH was determined at natural pH of each solution.

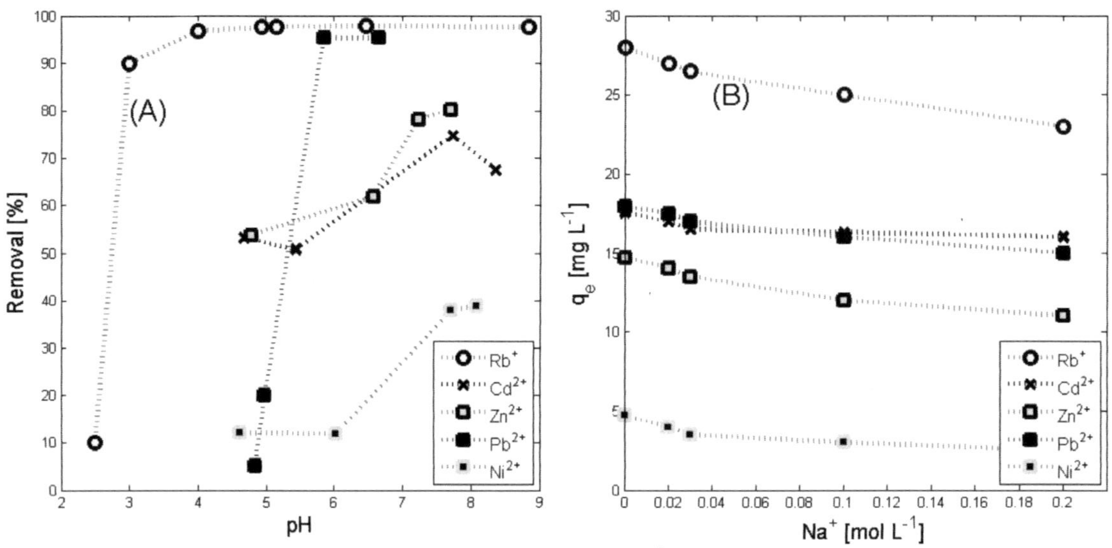

Figure 2. Effect of (A) pH (B) ionic strength (initial concentration: 100 mg L^{-1}).

3.3 Effects of ionic strength

The effect of ionic strength was evaluated by adding five different concentrations of $NaNO_3$ at 100 mg L^{-1} of initial metal concentration; see figure 2B. $NaNO_3$ as an inorganic salt plays an important roles in aqueous solution. The adsorption of metal ion on PmCH was decreased with a raise of $NaNO_3$ concentration due to performance of salt screens as an electrostatic interaction factor of the opposite charges between metal ions and PmCH surface [1].

3.4 Effects of contact time and adsorption kinetics

The effect of contact time between each ion and PmCH was studied at 3.5 g L^{-1} PmCH dosage and 100 mg L^{-1} ion concentration, which illustrated that the adsorption of ion was as a function of contact time. The pseudo-first-order (equation 2) and the pseudo-second-order kinetic (equation 3) models have been applied to determine and describe the kinetic process of liquid-solid phase [2]:

$$\ln(q_e - q_t) = \ln q_e - k_1 t \quad (2)$$

$$(t/q_t) = (1/k_2 q_e^2) + (t/q_e) \quad ; \quad h = k_2 q_e^2 \quad (3)$$

Studies showed that more than 90% of the total adsorption occurs within the first 100-150 min and the concentration approached their equilibrium value at 200-400 min, which can be confirmed by h value in Table 1. The initial adsorbent rate of Rb^+ was clearly more than the others. By comparing the correlation coefficient R^2, it appears that the pseudo-second-order kinetic fits the experimental data more accurately than the pseudo-second-order kinetic. This model suggests that ions are exchanged with hydrogen on the PmCH surface [3].

3.5 Equilibrium studies in single-component system

The effect of initial concentrations of each ion was carried out using Langmuir (equation 4) and Freundlich (equation 5) models within range of 10-100 mg L^{-1} of ions at 25°C [5]:

$$(C_e/q_e) = (1/q_{max} k_L) + (C_e/q_{max}) \quad (4)$$

$$\log q_e = \log k_f + (1/n)\log C_e \quad (5)$$

The adsorption capacities were increased with increasing

Table 1. The kinetics parameters for adsorption of ions using PmCH.

Ion	q_e [mg g^{-1}]	Pseudo-second-order			Pseudo-first-order	
		k_2 [g mg^{-1} min^{-1}]	h [mg g^{-1} min^{-1}]	R^2	k_1 [min^{-1}]	R^2
Rb^+	96.15	0.0008	0.6803	0.997	0.0076	0.903
Cd^{2+}	27.17	0.0005	0.1786	0.979	0.0070	0.980
Zn^{2+}	17.85	0.0007	0.1786	0.989	0.0047	0.971
Pb^{2+}	19.15	0.0009	0.3199	0.995	0.0041	0.949
Ni^{2+}	4.76	0.0015	0.0381	0.984	0.0068	0.990

the initial concentration of each ion. q_{max}, k_L, k_f, and n parameters were calculated from the slope and intercept of the above equations plots. The correlation coefficient (R^2) of Langmuir model (for all of ions were in the range of 0.992-0.998) fits the experimental data better than that of Freundlich isotherm (for all of ions were in the range of 0.819-0.965) which indicates that PmCH has a homogeneous surface, with equal energy in its activate sites. The maximum adsorption capacities were 96.15, 27.17, 17.85, 19.15, and 4.76 mg g^{-1} of Rb^+, Cd^{2+}, Zn^{2+}, Pb^{2+}, and Ni^{2+}, respectively. The k_L parameters were 0.176, 0.064, 0.110, 0.515, and 0.212 L mg^{-1} of Rb^+, Cd^{2+}, Zn^{2+}, Pb^{2+}, and Ni^{2+}, respectively.

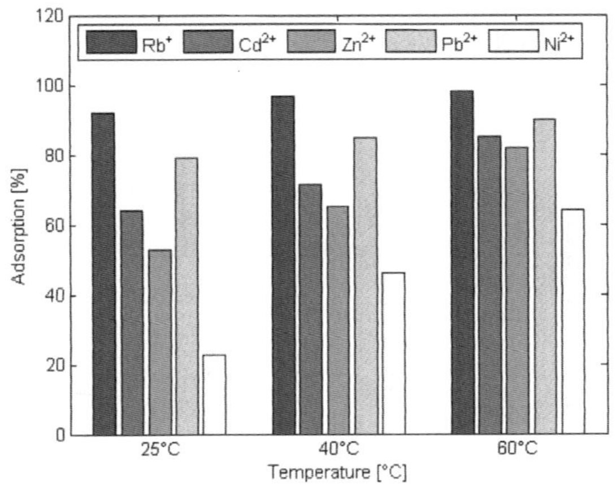

Figure 3. Effect of temperature at 25, 40, and 60°C (initial concentration: 100 mg L^{-1}).

3.6 Effects of temperature

The effects of temperature on the adsorption of heavy metals onto PmCH were evaluated at three different temperatures (25, 40, and 60 °C) and 100 mg L^{-1} of initial metal concentration; see figure 3. The adsorption capacity of each heavy metal onto PmCH was increased with increasing the temperature, which indicates that the adsorption of heavy metals onto PmCH is endothermic process.

Enhance of adsorption capacity with increasing the temperature is due to the strengthening of adsorptive forces between the adjacent molecules of the adsorbed phase, and between the ion species and active sites of the PmCH.

3.7 Multi-component adsorption (in binary mixtures)

To determine the competitive effect of ions on each other in binary mixture, removal percentage of 1st ion are shown in Table 2 at 3.5 g L^{-1} PmcH dosage and 100 mg L^{-1} ion concentration. The removal percentage of Rb^+ (as the 1st ion) in the presence of Cd^{2+}, Zn^{2+}, Pb^{2+}, and Ni^{2+} (as the 2nd ion) were 84.3, 89.7, 84.1, and 91.5 percent. However, the removal percentage of Cd^{2+}, Zn^{2+}, Pb^{2+}, and Ni^{2+} (as the 1st ion) in the presence of Rb^+ (as the 2nd ion) were 32.1, 22.1, 46.7, and 9.3 percent. These results illustrate that Rb^+ was more to be adsorbed onto PmCH even in binary mixture. The rank of removal efficiency in binary mixtures was $Rb^+ > Pb^{2+} > Cd^{2+} > Zn^{2+} > Ni^{2+}$ which can be explained by Lewis acids low and classification of heavy metals in metal into class-B 'soft ions', class-A 'hard ions' and borderline ions [1]. Rb^+, Cd^{2+}, Pb^{2+}, Zn^{2+} and Ni^{2+} on the basis of the Lewis acids low are classified in A, B, B, borderline, and borderline, respectively [6].

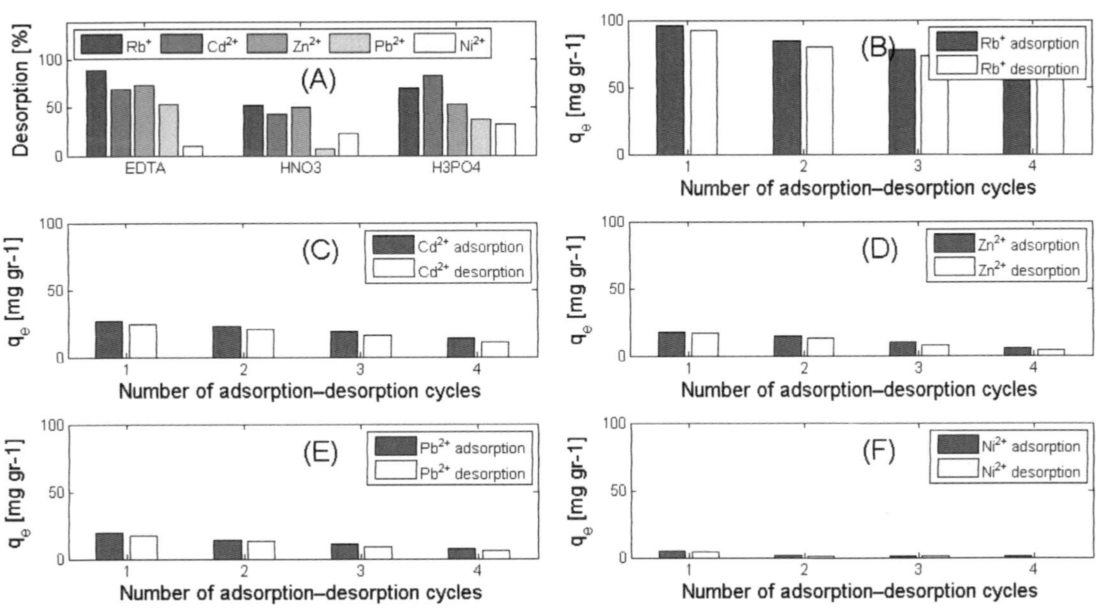

Figure 4. Effect (A) Desorption percentage of ions from PmCH using three different buffers. Adsorption–desorption cycles of PmCH for (B) Rb^+-EDTA, (C) Cd^{2+}-H_3PO_4, (D) Zn^{2+}-EDTA, (E) Pb^{2+}-EDTA, and (F) Ni^{2+}-H_3PO_4.

Table 2. PmCH Removal percentage based on 1st ion removal in binary mixtures at 25°C.

1st ion	2nd ion				
	Rb^+	Cd^{2+}	Zn^{2+}	Pb^{2+}	Ni^{2+}
Rb^+	-	84.3	89.7	84.1	91.5
Cd^{2+}	32.1	-	54.4	48.0	59.3
Zn^{2+}	22.1	39.4	-	36.5	48.2
Pb^{2+}	46.7	53.5	68.2	-	75.3
Ni^{2+}	9.3	9.9	16.4	6.8	-

3.8 Desorption and reusability

Desorption studies of metal ions were carried out to evaluate the reusability of PmCH using three different buffers (EDTA, HNO_3, and H_3PO_4); see figure 4. The best buffers for excellent desorption efficiency for Rb^+, Cd^{2+}, Zn^{2+}, Pb^{2+}, and Ni^{2+} ions were EDTA, H_3PO_4, EDTA, EDTA, H_3PO_4, respectively (figure 4A). The coordination ligands of each buffer formed a strong bonding with a selective ion, when each buffer was added to the ion adsorbed-PmCH. Ion–ligand bonding Strength was the determining factor for making more easily desorb from the PmCH. The reusability of PmCH using the selective buffer for each ion by following four cycles of adsorption–desorption process was shown in figure 4B-F. The negligible changes of adsorption–desorption cycles indicate that PmCH is an adsorbent with no irreversible sites and can be used as a reusable adsorbent in wastewater treatment.

4 Conclusions

PmCH was synthesized and characterized for removal of heavy metals by FTIR and XRD. The adsorption of metal ion was decreased with a raise of $NaNO_3$ concentration. The Langmuir isotherm and pseudo-second-order kinetic models fit the experimental data better than the others. The maximum adsorption capacities were 96.15, 27.17, 17.85, 19.15, and 4.76 mg g^{-1} of Rb^+, Cd^{2+}, Zn^{2+}, Pb^{2+}, and Ni^{2+}, respectively. The adsorption capacity of each ion was increased with increasing the temperature. The rank of removal efficiency in binary mixtures was $Rb^+ > Pb^{2+} > Cd^{2+} > Zn^{2+} > Ni^{2+}$. The best buffers for excellent desorption efficiency for Rb^+, Cd^{2+}, Zn^{2+}, Pb^{2+}, and Ni^{2+} ions were EDTA, H_3PO_4, EDTA, EDTA, H_3PO_4, respectively.

References

1. A.Z. Badruddoza, Z.B. Shawon, W.J. Tay, K. Hidajat, M.S. Uddin, Carbohydr. Polym. **91**, 1 (2013)
2. M.A. Moosavian, N. Moazezi, Desalin. Water. Treat. **57**, 44 (2016)
3. N. Moazezi, M.A. Moosavian, J. Environ. Chem. Eng. **4**, 2 (2016)
4. H. Li, Q. Gao, L. Chen, W. Hao, Sensors Actuators, B Chem. **173**, (2012)
5. T. Yousefi, M. Torab-Mostaedi, M.A. Moosavian, H. Ghasemi Mobtaker, Prog. Nucl. Energy **85**, (2015)
6. L. Yang, S. Wu, J.P. Chen, Ind. Eng. Chem. Res. **46**, (2007)

https://doi.org/10.1051/e3sconf/20185103005

Evaluation of the shoreline effect of the marine energy farms in different coastal environments

Eugen Rusu[1,*] and Florin Onea[1]

[1]Department of Mechanical Engineering, ''Dunarea de Jos'' University of Galati, Domneasca" Street, 47, Galati 800008, Romania

Abstract. The objective of the proposed work is to assess the possible effects on the shoreline dynamics of the marine energy farms. Three different coastal environments have been considered as case studies. The first area is located in the Portuguese continental nearshore at the European western coast of the Atlantic Ocean. The second in the Mediterranean Sea in the coastal environment of Sardinia Island and the third on the western side of the Black Sea in the Romanian nearshore. Besides the fact that, from a geographical point of view, the three areas targeted are located in very different zones, they represent also coastal environments with very different characteristics. A computational framework joining a spectral phase averaged wave model with a 1D parametric circulation model has been used. The numerical models are SWAN (acronym from Simulating Waves Nearshore) for the waves and NSSM (the Navy Standard Surf Model) for estimating the longshore currents. For each case study, the main characteristics of the environmental matrix have been analyzed first. Then, various transmission scenarios have been, designed considering also in each case different distances from the marine energy farm to the shoreline. A general conclusion of the proposed work is that the longshore currents are very sensitive to the presence of the marine energy farms and that is why such farms can also play an active role in coastal protection. Moreover, since it presents in parallel similar analyses for three coastal environments that have very different features, the present study provides a more comprehensive picture concerning the medium to long term impact on the shoreline dynamics of the future marine energy farms operating in the nearshore.

1 Introduction

The coastal areas are continuously shaped by the erosion and accretion processes which occur on various temporal and spatial scales. Although, these are common events which result from the natural and anthropogenic factors combination [1], during the recent years the beach erosion areas was significantly intensified, being estimated that only in Europe almost 15 km^2 are lost into the sea (per year in 2010). In order to attenuate such coastal hazards, nearly €3 billion/year is allocated for this issue on a European level [2, 3].

The wave action represents one of the most aggressive factors, if we consider that most of the wave energy dissipates in the surf area and the mechanical abrasion of the waves dislocate the sediment particles and carry them into the offshore areas. The breaking waves generate the longhsore currents which accelerate the erosion processes by transporting the particles in suspension along the coastline. The wave energy and longshore velocity significantly increase during the storm events, the beach cells being unable to regenerate in a reasonable period of time, especially if they show irregular features [4]. Various coastal protection systems emerged during time in order to reduce the coastal erosion, which include soft solutions (beach nourishment) or hard engineering, such as: sea walls and dikes, revetments or breakwaters [5]. These solutions require a high budget and may be considered to have limited efficiency, since they act on the effects of the erosion and do not tackle the source of the problem, which is the wave action [6].

In this case, a viable alternative comes from the renewable industry, more precisely from the wind and wave resources which are green energy and represent a sustainable source [7-9]. In terms of the coastal protection, better results are expected from the wave energy converters (WECs) developments which are capable to capture the wave energy from the offshore and nearshore areas. By deploying these generators on a wave farm configuration, it is possible to reduce more efficiently the wave heights and to avoid the regenerating effects of the wave fields, which may occur in the case of a single line breakwater.

Since most of the WECs include a connection to the seabed throughout a mooring system, this brings the opportunity to relocate the systems to other coastal sectors (if needed). Motivated by these aspects, the purpose of the present work will be to provide some insights about the effectiveness of a generic wave farm operating in various coastal environments.

2 Methods and materials

* Corresponding author: erusu@ugal.ro

The wave evolution in the geographical space will be assessed by using the SWAN (Simulating Waves Nearshore) wave model, which represents a state-of-the-art modeling tool based on the spectrum concept. This is capable to identify the sea state variation throughout the action balance equation, which is defined as [10]:

$$\frac{\partial N}{\partial t}+\nabla \vec{x}\cdot[(\vec{c}_g+\vec{U})\cdot N]+\frac{\partial C_\sigma N}{\partial_\sigma}+\frac{\partial C_\theta N}{\partial_\theta}=\frac{S_{tot}}{\sigma} \quad (1)$$

where, the left side represents the kinematic part of the equation, while on the right side the source and sink terms are given. These are associated with the most relevant physical processes. The waves and longshore currents variation are assessed throughout the ISSM (Interface for SWAN and Surf Models) which combines the models SWAN with the NSSM (Navy Standard Surf Model) [11].

Three different target areas were considered for evaluation, as illustrated in Figure 1, where the SWAN computational domains are presented. An important objective of the current work will be to identify the generic farm influence taking into account various distances from the shore, in this case being considered two scenarios: A) farm at ≈1.5 km (denoted with F1.5) and B) farm at ≈3 km (F3).

The first case study is in the Portuguese coastal environment, more precisely close to the Leixoes harbor, north to the city of Porto [12]. A maximum depth of 50 m defines the left corner of this area, which covers a surface of 36 km^2. A detailed evaluation of the wave variation in the vicinity of the shoreline comes from the two reference lines (L1 and L2) and throughout the four reference points (P1, P2, P3 and P4). A similar evaluation scheme will be used for all the areas taken into account. Going from an oceanic environment to an enclosed sea, Figure 1c presents a coastal sector close to the Sardinia Island, more exactly on the western side in the vicinity of Porto Ferro [13]. A bay defines this region, where the water depth does not exceed 100 m. The Black Sea environment was also considered in Figure 1d, in this case being selected a sector in the Romanian nearshore (Mangalia) which is located at the southern extremity of this region [14]. This computational domain is defined by a 5 by 5 km length in x and y direction, reporting a maximum depth of 25 m and a curbed shoreline configuration.

The numerical simulations involve several absorption scenarios. The first one is associated to a moderate absorption (denoted with MA), which corresponds to a transmission coefficient of 0.8 and a reflection coefficient of 0.05 [12]. This means that the WEC line absorbs only 20 % of the waves, while 5% of the waves are reflected. The second case study involves a highly absorbing configuration (*Ctr*=0.6; *Cr*=0.08), which simulates an ideal scenario, since almost 40% of the incoming waves will be available for the power generation.

Each geographical area is defined by a particular wave climate, therefore we need to consider a reliable source of data to initiate the SWAN simulation. In this study it was considered the reanalysis wave database provided by the European Centre for Medium-range Weather Forecast [15], which corresponds to the 10-year interval 2007-2016. The most relevant wave parameters (*Hs*–significant wave height, *Tm*–wave period; *Dir*– wave direction) were processed from a grid defined by a spatial resolution of 0.75°×0.75° and considering a time resolution of four data per day (00-06-12-18 UTC). Table 1 presents the most relevant wave patterns, indicated in terms of the average and extreme values.

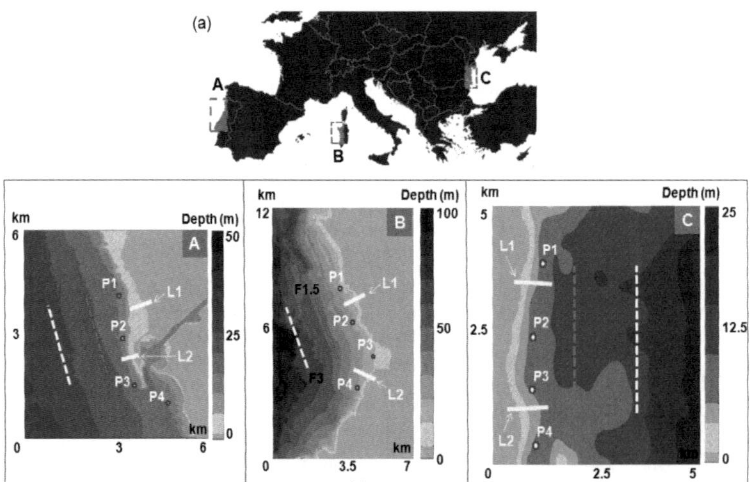

Figure 1. (a) Location of the three target areas and the computational domains considered for the SWAN simulations (b) Leixoes in the Portuguese nearshore, (c) Porto Ferro in the coastal environment of Sardinia Island and (c) Mangalia in the Romanian nearshore). In the background the bathymetry is presented while in the foreground the set-up of the wave farm scenario and the reference lines and points selected for the analysis.

Table 1. Extreme wave conditions, according to the ECMWF data (time interval 2007-2016) in the vicinity of the three target areas (Leixoes located in the Portuguese nearshore, Porto Ferro in the coastal environment of Sardinia Island and Mangalia located in the Romanian nearshore).

Site	Conditions	Wave parameters			SWAN scenarios
		Hs (m)	*Tm* (s)	*Dir* (°)	
Leixoes	Average	1.6	8.1	288	5 m – 13.6 s
	Extreme	5	13.6	-	– 288°
Porto Ferro	Average	0.9	4.9	232	5.4 m – 9.5 s
	Extreme	5.4	9.5	-	– 232°
Mangalia	Average	0.6	4.1	148	3.5 m – 7.7 s
	Extreme	3.5	7.7	-	– 148°

As expected, the Leixoes site presents more consistent values indicating an average significant wave height of 1.6 m and a maximum value of 5 m, with the mention that the Porto Ferro conditions present a maximum of 5.4 m while the average value is 0.9 m.

For the Mangalia site the most common heights vary around the value of 0.6 m, while during the extreme events a maximum of 3.5 m may occur. In terms of the wave direction, we observe that each target area is defined by a statistical value, which reflects the shoreline orientation and the direction from which the waves may occur. Taking into account that during the extreme events (ex: storms), the beach sectors are more affected by the erosion processes, only these extreme conditions will be further considered for evaluation (see Table 1).

3 Results

A first evaluation of the WEC influence is presented in Figure 2, considering also the scenario when both WEC lines (F1.5+F3) will operate, as in the case of a wave farm configuration. For the no farm situation, it can be observed the distribution of the wave fields and the tendency of the waves to rotate as they enter in the surf area. The shielding effect induced by the farm is influenced by the line position according to the incident waves and shoreline orientation, being also influenced by the distance to the shore. In the case of the F3 scenario, the values from the interval 4-4.5 m seems to occur mainly in the southern part of the Leixoes harbor compared to F1.5 scenario.

The absorption property of the farm also plays a key role, which is revealed by the multiple wave fields that occur between the farm and the shoreline. The presence of the two WEC lines (MA set-up) seems to replicate the effects observed for the study F3-HA, with the mention that the shielding effect is noticed closer to the shoreline.

Figure 3 illustrates a similar distribution, considering this time for the western part of the Mediterranean Sea, where the waves are coming from the south-west sector, and more consistent variations occur in the case of the HA scenario. Figure 4 is focused on the Mangalia region, where in general the central part reports wave conditions of 2-3 m compared to the lower-right corner where maximum heights of 4 m may occur.

For this environment, the F1.5 WEC line influence seems to count more. From the maps representing the spatial variations of the wave height we can estimate the size of the shielding and the beach sectors which is influenced, but it is difficult to estimate the intensity of this effect. Table 2 illustrates the *Hs* variations for the no farm situation (in meters) and for the WEC influence.

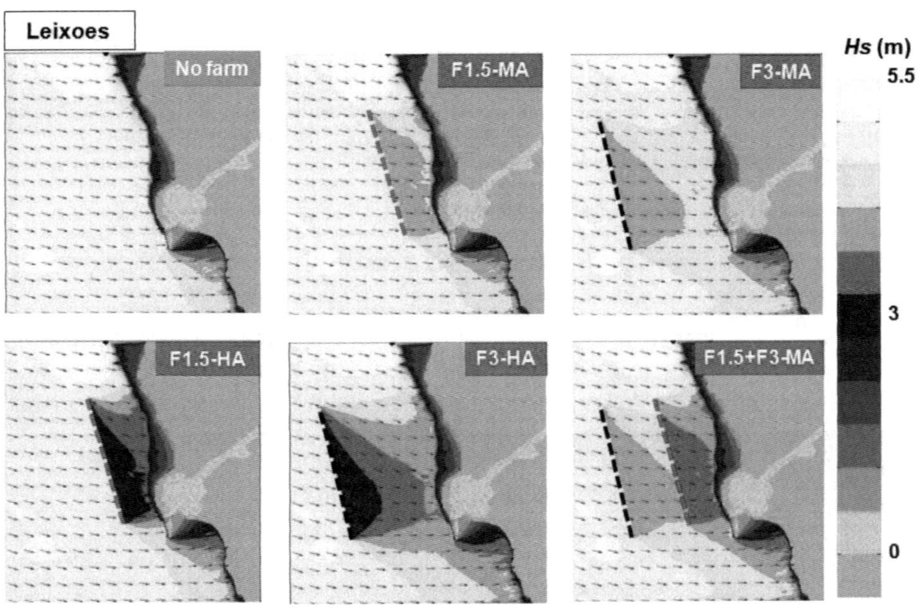

Figure 2. Evaluation in the geographical space (significant wave height scalar fields and wave vectors) based on the influence of the generic farm operating in the Leixoes region (Portugal) under extreme conditions.

According to these variations (in %), the significant wave heights will be reduced, regardless the case studies considered for evaluation. For the Leixoes area, more important values occur in the vicinity of the reference points P1 and P2, in the case of the F1.5-HA scenario (max 35.93%). The two-line configuration may represent a second option for the coastal protection revealing a maximum of 28.2% close to the point P2. From the MA results, it seems that a farm located at 1.5 km will be more effective than a project located in the offshore region.

The Sardinia Island case study (Porto Ferro) reveals a similar trend, where a maximum attenuation of 37.3% is accounted by the F1.5 configuration. A two-line wave farm may reduce the waves with 27.57%, being followed by the F1.5 line with 19.14%. For the Mangalia region, which represents a marine environment defined by moderate wave resources, seems that the best protection is provided by a farm located close to the shore (at 1.5 km). Another viable solution will be to consider a two-line wave farm, which may attenuate the waves between the reference points P1 and P3 with almost 10%.

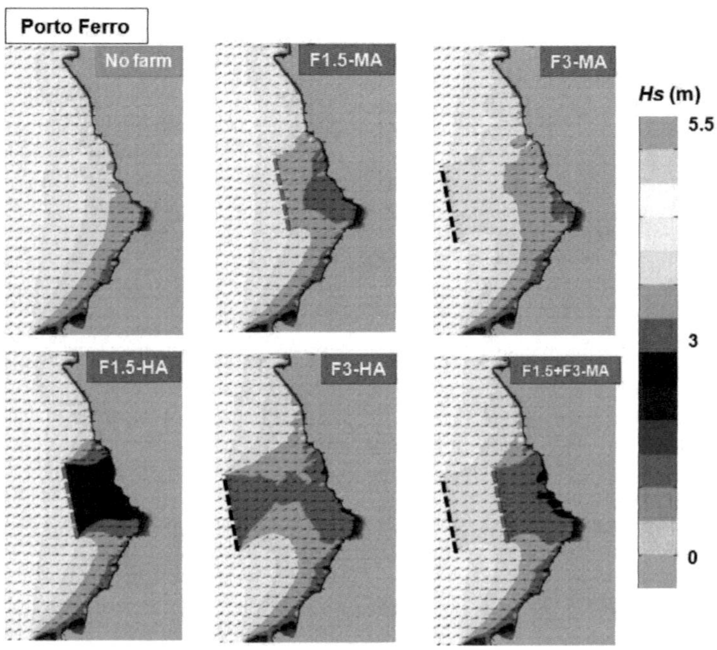

Figure 3. Evaluation in the geographical space (significant wave height scalar fields and wave vectors) based on the influence of the generic farm operating in the Porto Ferro region (Sardinia Island) under extreme conditions.

Table 2. Variation of the *Hs* values (in %) in the presence of the wave farm, where: F1.5-wave farm located at 1.5 km from the shore; F3-wave farm located at 3 km from the shore; MA - moderate absorption; HA-high absorption. With **bold font** the values higher than 10%, are indicated while with underline font the values exceeding 20% are marked.

Site		No farm (m)	F1.5-MA (%)	F3-MA (%)	F1.5-HA (%)	F3-HA (%)	Two lines-MA (%)
Leixoes	P1	4.71	14.5	9.47	28.44	**18.3**	22.32
	P2	4.59	**17.49**	**10.69**	35.93	21.96	28.2
	P3	4.65	6.41	9.60	**10.23**	**17.12**	**14.53**
	P4	3.82	1.59	5.43	2.431	**10.18**	6.65
Porto Ferro	P1	4.26	7.53	9.16	**13.72**	**16.42**	**15.25**
	P2	4.25	**19.14**	**10.9**	37.3	20.11	27.57
	P3	3.73	**10.12**	7.30	**18.84**	**13.53**	**15.37**
	P4	3.78	1.25	4.60	2.091	8.161	5.454
Mangalia	P1	1.85	5.48	3.36	**17.33**	8.642	**11.12**
	P2	2.02	7.56	4.14	22.2	9.912	**13.7**
	P3	1.84	5.97	2.32	**14.35**	6.537	**10.01**
	P4	1.08	1.44	1.02	4.227	2.642	3.162

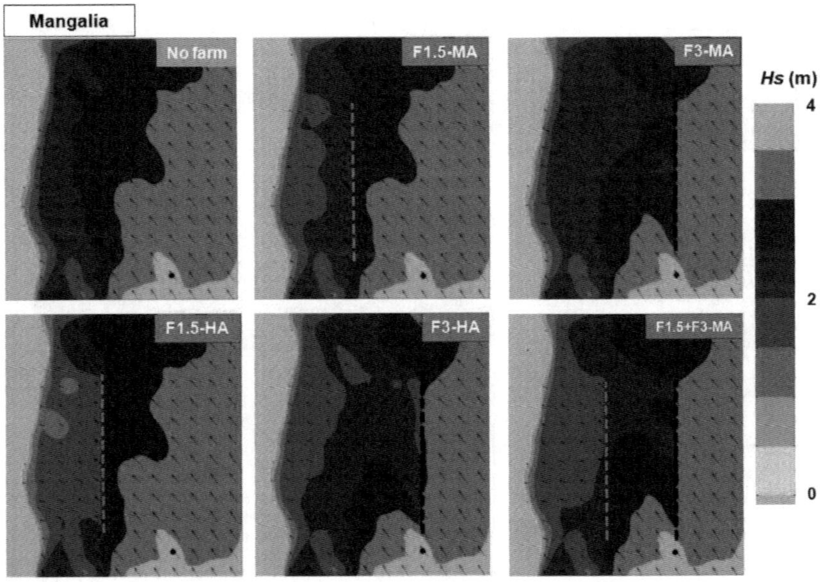

Figure 4. Evaluation in the geographical space (significant wave height scalar fields and wave vectors) based on the influence of the generic farm operating in the Mangalia region (Romanian nearshore) under extreme conditions.

Besides the significant wave height, the wave direction represents another important factor for the coastal erosion, since indicates the direction from where the beach sector will be more vulnerable. In the presence of an obstacle (such as the wave farm is) the initial direction may be shifted, and this variation is illustrated in Table 3. The variations reveal positive or negative values, reporting this time more important value for the Mangalia sector, where a maximum of 5.88 may occur close to the reference point P3 in the case of the F1.5-HA. In general, higher fluctuations may occur in the case of the two-line configuration with high absorption.

In the vicinity of the coastline, the longshore currents may change the balance between the accretion and erosion processes by influencing the flux of the sediment transport. In this respect, Figure 5 illustrates the profile of the longshore currents along the reference lines L1 and L2, from which it can be observed that in some cases (L1-Porto Ferro and L1-Mangalia) the presence of the WEC lines reduces the distance to the shore. For example, in the case of Mangalia (no farm situation) the profile reveals a total length of 400 m, which may decrease until 250 m in the F1.5-HA. For the Leixoes region, we can notice that the WEC influence increases the current velocity from 0.5 m/s (no farm situation) to a value located close to 2 m/s (F3-HA).

Table 4 presents the maximum current velocity, for all the areas targeted and the reference lines considered. In the case of the case studies Leixoes and Porto Ferro, we can notice that for the no farm scenario the current velocity is insignificant ($\approx$ 0 m/s), gradually increasing in magnitude according to the selected WEC configuration. Although Porto Ferro and Mangalia represent semi-enclosed basins, they indicate much higher value for the velocity of the nearshore currents than in the case of the Leixoes area, indicating maximum of 1.74 m/s and 1.29 m/s respectively.

Table 3. Variation of the mean wave direction (*Dir* parameter in %), where with the **bold font** the negative variations in relationship with the no farm situation are indicated.

Site		No farm (°)	F1.5–MA (%)	F3–MA (%)	F1.5–HA (%)	F3–HA (%)	Two lines-MA (%)
Leixoes	P1	272.5	**-1**	**-1.32**	**-2.67**	**-2.90**	**-2.39**
	P2	275.2	0.29	**-0.63**	**-0.11**	**-1.72**	**-0.38**
	P3	265.5	1.48	1.42	2.45	2.74	2.87
	P4	257.3	0.37	1.10	0.56	2	1.43
Porto Ferro	P1	240.8	**-1.48**	**-0.95**	**-3.01**	**-1.93**	**-2.50**
	P2	250.1	**-0.22**	0.35	**-0.56**	0.83	0.09
	P3	260.4	1.06	0.71	2.29	1.48	1.76
	P4	258.2	0.37	1	0.63	1.85	1.28
Mangalia	P1	112.3	0.61	**-1.06**	1.77	**-2.06**	0.14
	P2	106.6	**-1.13**	**-2.33**	**-3.34**	**-5.09**	**-3.41**
	P3	91.92	**-1.82**	**-1.24**	**-5.88**	**-2.01**	**-2.77**
	P4	110.9	0.09	**-1.12**	0.25	**-2.37**	**-0.72**

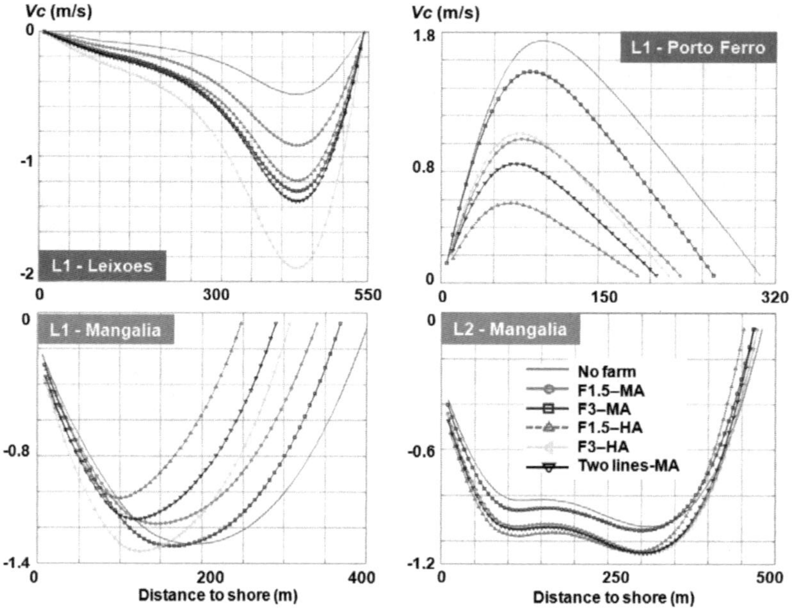

Figure 5. Nearshore current profiles indicating the impact of the generic wave farm in the vicinity of: line L1-Leixoes; line L2-Porto Ferro; lines L1 and L2 - Mangalia.

Table 4. Values of the maximum current velocities (*Vcmax* in m/s) reported in the presence of the wave farms in the three target areas taking into account.

Site		No farm	F1.5–MA	F3–MA	F1.5–HA	F3–HA	Two lines-MA
Leixoes	L1	0.5	0.91	1.28	1.19	1.9	1.36
	L2	0	0.36	0.22	0.6	0.38	0.58
Porto Ferro	L1	1.74	1.03	1.52	0.58	1.07	0.85
	L2	0	0.22	0.24	0.5	0.53	0.51
Mangalia	L1	1.29	1.18	1.30	1.03	1.33	1.15
	L2	0.94	1.05	0.95	1.04	1.05	1.05

4 Conclusions

In the present work some insights were provided regarding the coastal variations induced by various WEC configurations which may operate in different coastal environments. For the Leixoes region (Portugal) it was noticed a significant attenuation of the waves, which is associated with an increase of the current velocity from zero to 0.58 m/s (line L2). For this region all the configurations defined by a high absorption property seem to be efficient, of which can be added the two-line configuration. A similar trend is observed in the Porto Ferro environment, with the mention that the current velocity decreases along the line L1 and gradually increases along the L2. For Mangalia, more consistent results may be reported in the case of a single WEC line located close to the shore (F1.5-HA), while a two line configuration may be also considered for implementation.

Finally, it can be mentioned that the coastal erosion represents a serious threat to the beach stability and the emerging of WEC technology may offer additional opportunities for the coastal protection.

Acknowledgements

This work was carried out in the framework of the research project REMARC (Renewable Energy extraction in MARine environment and its Coastal impact), supported by the Romanian Executive Agency for Higher Education, Research, Development and Innovation Funding – UEFISCDI, grant number PN-III-P4-IDPCE-2016-0017. The ERA-Interim data used in this study have been obtained from the ECMWF data server.

References

1. M. Lopez, J.L. Pagan, I. Lopez, L. Aragones, A.J. Tenza-Abril, J. Garcia-Barba, Factors influencing the retreat of the coastline. International Journal of Computational Methods and Experimental Measurements **5**, 741-649 (2017).

2. CONSCIENCE, Coastal erosion control based on the concept of sediment cells. Deltares, the Netherlands (2010).
3. EUROSION, A guide to coastal erosion management practices in Europe: lessons learned. *National Institute of Coastal and Marine Management of the Netherlands* (2004).
4. R. Davidson-Arnott, An introduction to coastal processes and geomorphology. Cambridge University Press (2010).
5. A.T. Williams, A. Giardino, E. Pranzini 2016. Canons of Coastal Engineering in the United Kingdom: Seawalls/Groynes, a Century of Change? Journal of Coastal Research **32-5**, 1196-1211 (2016).
6. EUROSION, Living with coastal erosion in Europe sediment and space for sustainability. European Commission, ISBN 92-894-7496-3 (2004).
7. E. Rusu, F. Onea, Estimation of the wave energy conversion efficiency in the Atlantic Ocean close to the European islands. Renewable Energy **85**, 687-703 (2016).
8. F. Onea, R. Alina, E. Rusu, Evaluation of the wind energy potential in the coastal environment of two enclosed seas. Advances in Meteorology, doi:10.1155/2015/808617 (2015).
9. L. Rusu, F. Onea, The performance of some state-of-the-art wave energy converters in locations with the worldwide highest wave power. Renewable and Sustainable Energy Reviews **75**, 1348-1362 (2017).
10. SWAN, Scientific and technical documentation (SWAN Cycle III version 41.01A). Delft University of Technology (2015).
11. D.C. Conley, E. Rusu, The middle way of surf modeling. In: Proceedings of the 30th International Conference on Coastal Engineering, ICCE 2006, 2-9 September. San Diego, USA: World Scientific Pub Co Inc, Coastal Engineering, 1053-1065 (2006).
12. F. Onea, E. Rusu, The expected efficiency and coastal impact of a hybrid energy farm operating in the Portuguese nearshore. Energy **97**, 411-423 (2016).
13. F. Onea, L. Rusu, Coastal impact of a hybrid marine farm operating close to Sardinia Island. OCEANS'15 MTS/IEEE GENOVA 18-21 May 2015 Genova, Italy (2015).
14. A.T. Zanopol, F. Onea, E. Rusu, Longshore curents evaluation along the Romanian Black Sea coast. 14th International Multidisciplinary Scientific Geoconference (SGEM) Location: Albena, Bulgaria Date: 17-26 June, **2**, 637-644 (2014).
15. European Centre for Medium-Range Weather Forecasts (ECMWF), IFS Documentation – Cy31r1, Operational implementation. Part VII: ECMWF wave mode. European Centre for Medium-Range Weather Forecasts, Shinfield Park, Reading, England (2006).

https://doi.org/10.1051/e3sconf/20185103006

An air circulation purifying system based on PLC

Yin Chen[1], Chengliang Han, Peng Chen, Zong Chen, L Y Liu and T Du [1,*]

[1]School of Metallurgy, Northeastern University, Shenyang, Liaoning Province, China

Abstract. Existing air purifiers are mostly indoor air circulation, and most of them cannot achieve indoor and outdoor air circulation and provide fresh air to room. At the same time, the current air purifier filter on the market after absorbing a large amount of dust and other pollutants, if not replaced in time, is likely to cause secondary pollution. In response to these problems, this paper carried out an exploratory research that can achieve indoor and outdoor air purification device. According to the clean air volume of HEPA filter, the pressure of fan and the cross-sectional area of filter are determined, meanwhile, the air resistance of indoor and outdoor circulating air purifier, the effect of air blower on air purification efficiency and filter CADR are discussed in detail. According to the filtering characteristics of single-layer filter, a kind of indoor and outdoor circulating air purifier-single-layer filter type air purifier was designed. Among the air purifier, the PLC control circuit is used to achieve single-layer filter air purifier filter automatically replace.

1 Introduction

In the winter of 2012, air pollution in China was so severe that one-third of the Chinese territory was covered by fog and haze, and the continuously hazy weather in most big cities lasted for more than 30 days [1]. At the beginning of 2016, haze appeared in many places in China. During the severe smog period, many highways were blocked; the aircraft were grounded; the school was closed; patients with respiratory diseases increased significantly, and the normal production and living and physical health of the local residents have been severely damaged. The main component of haze is PM2.5. Research shows exposure to particulate matter less than 2.5 μm in diameter (PM2.5) is associated with asthma and respiratory symptoms and etc [2].

It should be pointed out that the annual average concentration of PM2.5 standard set by the World Health Organization is less than 10 μg / m3, and the annual average concentration of 35 μg / m3 will greatly increase the incidence of haze and death. A recent Chinese case study concluded that the reduction in life expectancy of about 3 years may be expected from long-term exposure to an additional 100 μg/m3 of Total Suspended Particles (TSPs) [3]. The status quo of haze pollution in China has been extremely serious. The problem of haze pollution control has long been impeded.

There are several purification technologies in air purifiers such as HEPA(high efficiency particulate air filter), mineralization technology of ultra-light structure, negative ion technology, the photocatalyst technology, electrostatic dust collection, activated carbon adsorption technology and etc [4].

There are two main types of air purifiers on the market in China. The first air purifier usually uses the filter with activated carbon to adsorb the tiny particles. The working capacity of this kind of purifier is limited. The purifying efficiency of the harmful gas and the dust will gradually decrease with the use time until it is completely deactivated. At this moment, the pollutants will accumulate in the indoor and the filter net, which seriously threatens human health. In severe haze, the filter needs to be replaced every 0.5-3 months, and each time the cost of more than 100 RMB, greatly increased the cost of late maintenance. The second air purifier uses high-voltage electrostatic to capture dust particles, but electrostatic dust removal technology can only remove particulate pollutants, cannot remove harmful gas pollutants. In addition, electrostatic dust collection has discharge electrode in which a direct current is applied [5], and the relevant high-voltage electrostatic generator, such as the discharge needles, need regular cleaning. The process is difficult to operate, and easily damaged, expensive, especially high voltage electrostatic technology will produce a large number of harmful ozone. At present, two types of air purifiers on the market are the circulation of the original indoor air, the circulation of indoor and outdoor air cannot be achieved, and the fresh air cannot be provided to the room.

2 Methodology

This paper designs an air purifier that can purify the outdoor air to the room and fresh air is supplied to the room by filtering outdoor air. The air pressure in the room is greater than the outdoor air pressure so that the air in the room passes through the gaps to the outdoor one-way circulation and prevents outdoor air backflow from bringing atmospheric pollutants into the room. In addition, the current air purifier filter on the market cannot be

*Corresponding author: dut@smm.neu.edu.cn

https://doi.org/10.1051/e3sconf/20185103006

automatically replaced. Once used for some time, the filter will accumulate large amounts of dust and other pollutants in the absorption. If not timely replacement, it will be likely to cause secondary pollution. Therefore, this paper designed a single-layer filter air purifier that can automatically replace the filter through the PLC. Compared to the folded filter, this single-layer filter design will greatly improve the filter's life and the volume of dust of the same area. The price will be relatively decreased, while not causing secondary pollution of indoor air.

2.1. HEPA filter

HEPA filter efficiency for the 0.3 micron fine particles can reach 99.7%. Its characteristic is the air can pass, but the tiny particles cannot pass. HEPA filter is the most effective smoke, dust, bacteria and other pollutants filter material. Therefore, this paper chose Spring CMI300 HEPA filter net as a filter material.

2.2. Differential pressure gauge

The Testo510 (shown in Figure 1) is a device for measuring differential pressure. It is commonly used for measuring differential pressure (such as measuring filter net permeability) and gas-liquid pressure,. The readings are accurate and reliable.

2.3. Anemometer

In this paper, Testo405-V1 anemometer is used to measure the outlet air velocity of the air purifier, and the clean air volume of the air purifier is obtained according to the outlet cross-sectional area. Testo405-V1 (shown in Figure 2) is a thermal anemometer. With a telescopic handle (up to 300mm extension), it measures wind velocity, air volume and temperature. It is suitable for detecting ventilation ducts, vents or unsealed windows and can read out the number directly, responsive, easy to operate.

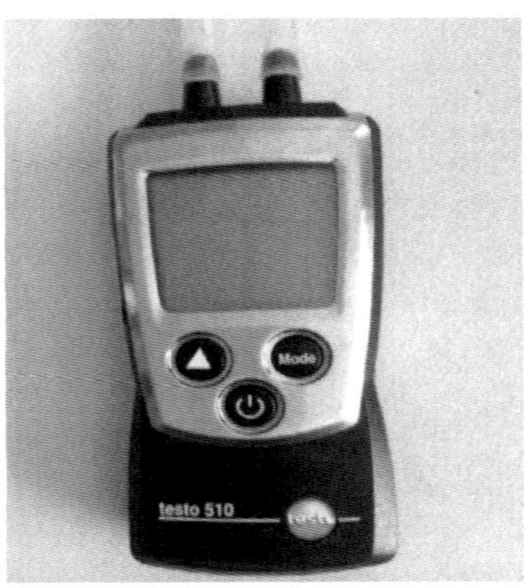

Figure 1. Differential pressure gauge(Testo510)

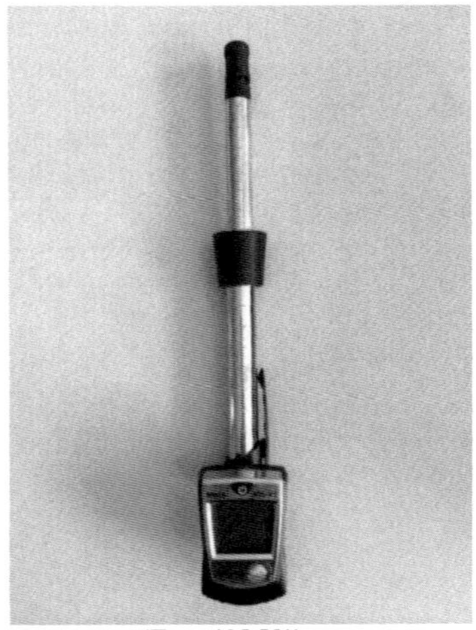

Figure 2. Anemometer(Testo405-V1)

2.4. Air purifier

In order to facilitate the study of the filter performance of the filter, the fan of Panasonic F-PDF35C air purifier was used as a new air purifier design of the fan, and the original filter net are replaced by the HEPA filter net we want to study. Then a series of performance parameters of the modified air purifier were tested to determine the PM2.5 filtration performance of our experimental raw HEPA filter. The Panasonic F-PDF35C air purifier (shown in Figure 3) with low, medium and high speed windshield design can achieve the expected wind speed.

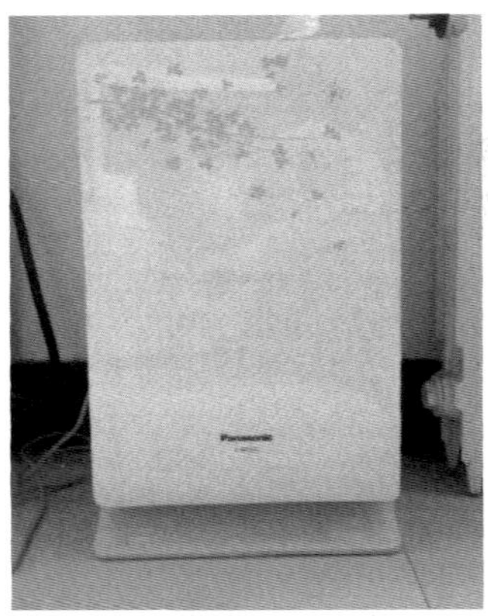

Figure 3. Air Purifier (Panasonic F-PDF35C)

3 Results and discussion

3.1. Pressure analysis of fan

Testo510-type pressure anemometer was used to test Panasonic F-PDF35C air purifier corresponding to the different levels of pressure, and that is the fan pressure under different wind velocity. When measuring, the Panasonic F-PDF35C air purifier does not have any HEPA filter installed. Open the air purifier power supply, respectively, in the low, medium and high three different wind speed levels, with a differential pressure test before and after the air purifier fan pressure. In order to reduce the experimental error, it was measured three times, and the final results take the average, the results are shown in Figure 4. It can be seen from Figure 4 that the differential pressures before and after the fan are 17Pa, 60Pa and 130Pa respectively at low wind velocity, middle speed velocity and high wind velocity, which provides the wind resistance for the design of the air purifier in the later chapters of this paper.

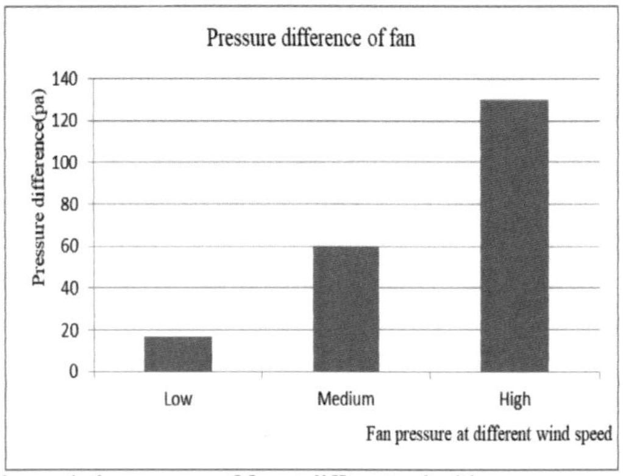

Figure 4. the pressure of fan at different velocities

3.2 Filtration efficiency analysis

In order to understand the filtration efficiency of the filter, the filtration efficiency of the PM2.5 in the aerosol of the environment under low, medium and high filtration velocities was tested. The specific experiment was carried out in a 60cm*60cm*60cm closed acrylic box. Lighting a cigarette, then using PM2.5 tester PM2.5 tests concentration in the acrylic box. When the PM2.5 concentration was 990 μg / m3, the air purifier was turned on and the initial concentration was set at 990 μg / m3. At this point, the outlet gas concentration of air purifier was regarded as the final concentration.

The filtration efficiency of the filter (η) is defined as the ratio of the difference between the initial concentration c1 of the air purifier outlet and the final concentration c2 to the initial concentration value. The calculation formula is shown in the following formula (1):

$$\eta = \frac{c_1 - c_2}{c_1} \times 100\% \qquad (1)$$

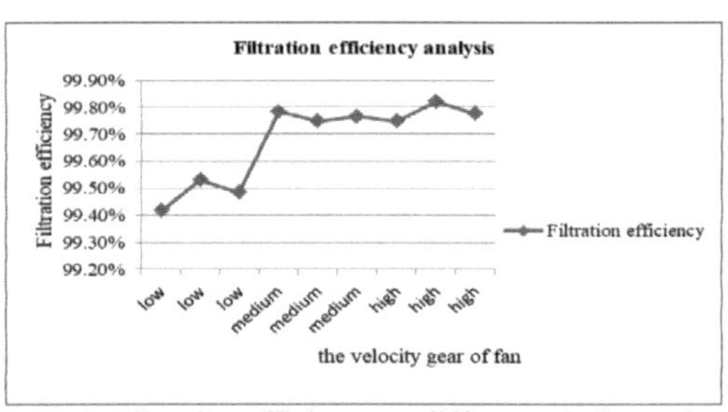

Figure 5. Filtration efficiency at different wind speed

The result of filtration efficiency analysis is shown in Figure 5. Regardless of the wind velocity of fan, the air purifier purification efficiency was always more than 99%, so the wind velocity had little effect on purification efficiency of the Spring CMI300 HEPA filter, and Spring CMI300 HEPA filter was suitable for air purifier filter.

3.3 CADR analysis

Air purifier outlet size: length × width = 11cm × 7cm. The outlet cross-sectional area is 77 cm2, and wind speed of the outlet can be measured by Testo405-V1 anemometer. The amount of clean air (CADR) can be calculated based on the filtration velocity (v) and the cross-sectional area (c). The calculation formula is shown in the following formula (2).

$$CADR = v \times c \qquad (2)$$

According to formula (2) to calculate the amount of clean air filter, the result is shown in Figure 6.

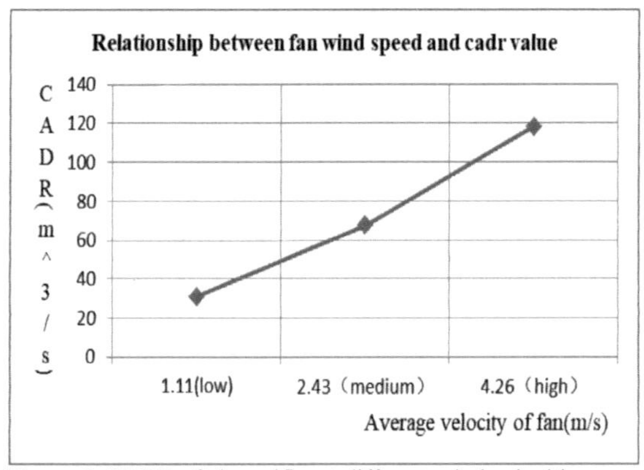

Figure 6. CADR of air purifier at different wind velocities

4 Model device

As is shown in Figure 7, the air purifier introduces the outdoor air into the room through a fan. After the filtration by the HEPA filter, clean purified air outside the room is introduced into the room to circulate indoor and outdoor air. When accumulative pollutants on the air purifier filter gradually increased, the purification effect will gradually decline. According to the consumption guideline of GB and AQSIQ, by reducing the CCM value (When the CADR value changes to 50% of the initial CADR value, the total weight of the purified pollutants in a unit of mg is the CCM value), CCM value can be converted into the average daily capacity. Then the number of filter days which is the filter replacement cycle can be calculated. As is shown in Figure 8 and Figure 9, when the filter net reaches the service life, the PLC control system starts the motor through the timer and automatically drives the filter net to be replaced. The process that filter net automatically changes the filter net is shown in Figure 10.

https://doi.org/10.1051/e3sconf/20185103006

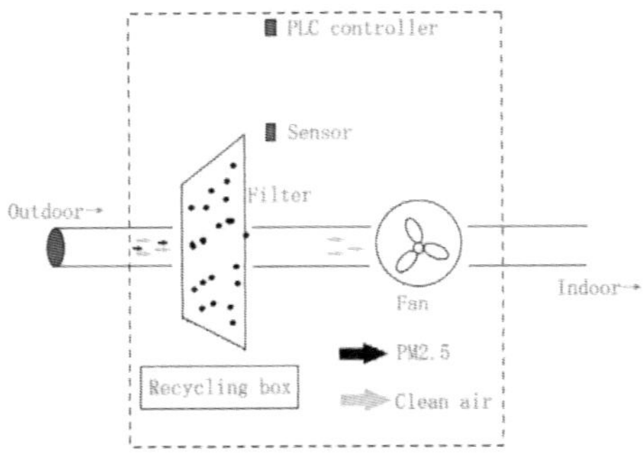

Figure 7. Adsorption process of filter net

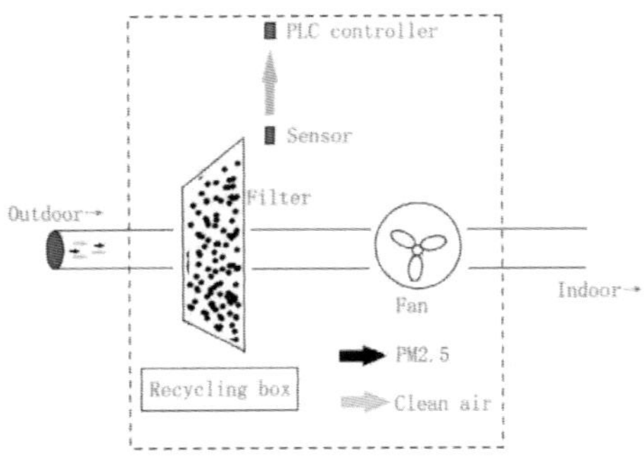

Figure 8. The filter is about to be full

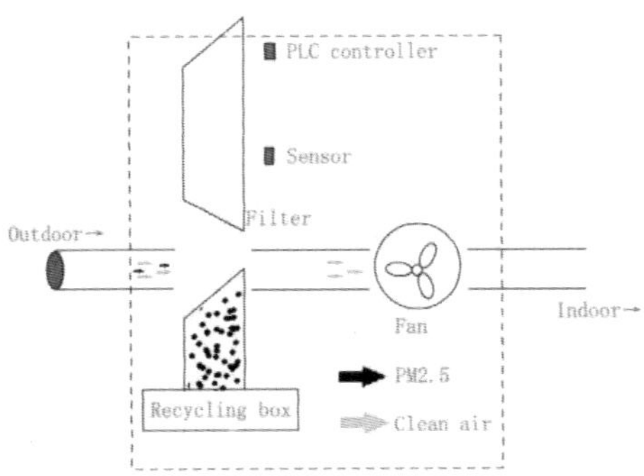

Figure 9. Automatic replacement starts

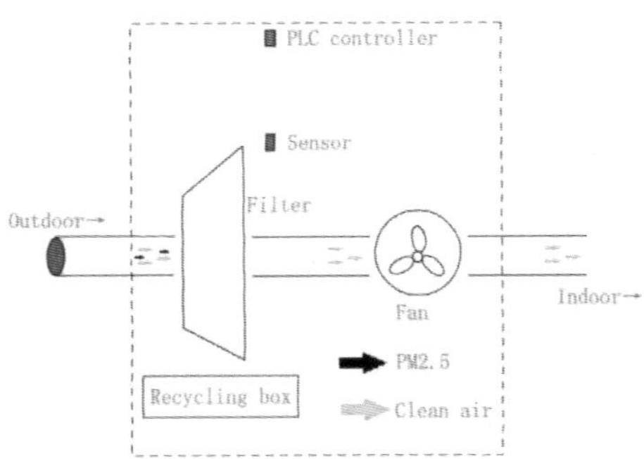

Figure 10. Automatic replacement ends

5 PLC control system

Programmable logic controller (PLC) is a basic and essential element in the engineering of industrial automation [6]. It uses a programmable memory to store instructions and execute functions including logical control, sequencing, timing, counting and arithmetic [7]. In this paper, the control system uses the Siemens S7-200 PLC with good reliability and scalability as the controller to design the LAD circuit. As is shown in Figure 11, I0.0 is the button, C0, C1, C2 is a counter, SM0.5 is a pulse signal, and Q0.0 is to replace the filter motor. The time base of this LAD circuit is one hour. VW is defined as replacement cycle in a unit of hour. When the filter reaches the end of its service life, Network 1 opens the way for the motor to start and automatic filter replacement is complete. Therefore, the problem of secondary air pollution has been solved.

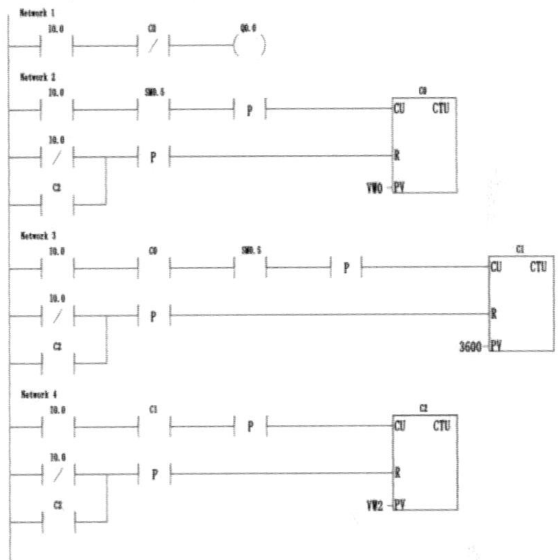

Figure 11. PLC automatic control system works

6 Conclusions

The current existing air purifier purifies the original air in the room, and cannot be indoor and outdoor air circulation and provide fresh air to the room. So, in this paper, the design of indoor and outdoor air purifier was studied. To sum up, first, the wind resistance, filtration speed, clean air volume and dust holding capacity of each filter were analyzed to find out the most suitable filter net for indoor and outdoor air purifiers. Then according to the characteristics of a single filter net, a kind of indoor and outdoor air purifier was designed.

As the air purifier filter cannot be automatically replaced in time, this paper uses the Siemens S7-200 series PLC programming software to achieve the air purifier replacement filter automatically. If system need to change the filter net replacement cycle, changing the value of VW can achieve it. Automatic replacement filter net of air purifier control system is only a typical application in the intelligent control model. With the rapid development of science and technology, the information technology with computer technology, communication technology and software technology as the core has achieved more rapid development. All kinds of equipment can be combined with this technology. Wide range of applications and computing systems, intelligent systems will gradually change the lives of humans.

Acknowledgments

The authors gratefully acknowledge the financial support of Innovation and entrepreneurship college of Northeastern University (Grants 170045) and Natural Science Foundation of China (Grants 51406029 and 51474067) and the guide of Professor T Du and Professor L Y Liu.

References

1. Chao Ye, Ruishan Chen, Mingxing Chen 2015 The impacts of Chinese Nian culture on air pollution, Sustainable Cities and Society, v 31, p 26-36, May 2017
2. Mimura, Tatsuya, Ichinose, Takamichi, Yamagami and Satoru 2014 Airborne particulate matter (PM2.5) and the prevalence of allergic conjunctivitis in Japan, Science of the Total Environment, v 487, n 1, p 493-499, July 15, 2014
3. Y. Chen, A. Ebenstein, M. Greenstone, H. Li
   Evidence on the impact of sustained exposure to air pollution on life expectancy from China's Huai River policy
   Proc. Natl. Acad. Sci. U. S. A., 110 (2013), pp. 12936-12941, 10.1073/pnas.1300018110
4. Yunhan Wang, Hao Wang, Changzhong Zhao, Yi Zhang 2015 Advanced Materials Research, v 1092-1093, 1025-8, 2015
5. SEKIKAWA T, ABE T, OGASAWARA M, SHIBAYAMA H 2016-61682LWO2016153046-A1; KR2017125968-A
6. M Mahadi, NA Mohd Ami, M Ab Rahim, MS Abdul Majid 2015 Applied Mechanics and Materials, v 786, 367-71, 2015
   DOI: 10.4028/www.scientific.net/AMM.786.367
7. C. D. Johnson, Process Control Instrumentation Technology, Prentice Hall., Upper SaddleRiver, New Jersey, 2006.

AUTHOR INDEX